Yvonne Scheidegger

Noble gas concentrations in stalagmite fluid inclusions

Yvonne Scheidegger

Noble gas concentrations in stalagmite fluid inclusions

A new climate proxy

Südwestdeutscher Verlag für Hochschulschriften

Impressum/Imprint (nur für Deutschland/only for Germany)
Bibliografische Information der Deutschen Nationalbibliothek: Die Deutsche Nationalbibliothek verzeichnet diese Publikation in der Deutschen Nationalbibliografie; detaillierte bibliografische Daten sind im Internet über http://dnb.d-nb.de abrufbar.
Alle in diesem Buch genannten Marken und Produktnamen unterliegen warenzeichen-, marken- oder patentrechtlichem Schutz bzw. sind Warenzeichen oder eingetragene Warenzeichen der jeweiligen Inhaber. Die Wiedergabe von Marken, Produktnamen, Gebrauchsnamen, Handelsnamen, Warenbezeichnungen u.s.w. in diesem Werk berechtigt auch ohne besondere Kennzeichnung nicht zu der Annahme, dass solche Namen im Sinne der Warenzeichen- und Markenschutzgesetzgebung als frei zu betrachten wären und daher von jedermann benutzt werden dürften.

Verlag: Südwestdeutscher Verlag für Hochschulschriften GmbH & Co. KG
Heinrich-Böcking-Str. 6-8, 66121 Saarbrücken, Deutschland
Telefon +49 681 37 20 271-1, Telefax +49 681 37 20 271-0
Email: info@svh-verlag.de

Approved by: Zürich, ETH Zürich, Dissertation Nr. 19547, 2011

Herstellung in Deutschland:
Schaltungsdienst Lange o.H.G., Berlin
Books on Demand GmbH, Norderstedt
Reha GmbH, Saarbrücken
Amazon Distribution GmbH, Leipzig
ISBN: 978-3-8381-3123-8

Imprint (only for USA, GB)
Bibliographic information published by the Deutsche Nationalbibliothek: The Deutsche Nationalbibliothek lists this publication in the Deutsche Nationalbibliografie; detailed bibliographic data are available in the Internet at http://dnb.d-nb.de.
Any brand names and product names mentioned in this book are subject to trademark, brand or patent protection and are trademarks or registered trademarks of their respective holders. The use of brand names, product names, common names, trade names, product descriptions etc. even without a particular marking in this works is in no way to be construed to mean that such names may be regarded as unrestricted in respect of trademark and brand protection legislation and could thus be used by anyone.

Publisher: Südwestdeutscher Verlag für Hochschulschriften GmbH & Co. KG
Heinrich-Böcking-Str. 6-8, 66121 Saarbrücken, Germany
Phone +49 681 37 20 271-1, Fax +49 681 37 20 271-0
Email: info@svh-verlag.de

Printed in the U.S.A.
Printed in the U.K. by (see last page)
ISBN: 978-3-8381-3123-8

Copyright © 2012 by the author and Südwestdeutscher Verlag für Hochschulschriften GmbH & Co. KG and licensors
All rights reserved. Saarbrücken 2012

Table of contents

1 Introduction 1
  1.1 Introduction and scope of work...1
  1.2 Outline 2

2 Scientific background...5
  2.1 Noble gases..5
    2.1.1 Noble gases in water as environmental tracers 5
    2.1.2 Determination of "noble gas temperatures (NGTs)" 7
  2.2 Stalagmites...8
    2.2.1 Stalagmites as climate archives 8
    2.2.2 Fluid inclusions in stalagmites 10
    2.2.3 Noble gas concentrations in fluid inclusions 10
  2.3 Analysis of stable isotopes 13
  2.4 Sample description...14

3 Accurate analysis of noble gas concentrations in small water samples and its application to fluid inclusions in stalagmites................................19
  3.1 Introduction 20
  3.2 Fluid inclusions in stalagmites...22
  3.3 Methods 22
    3.3.1 Extraction of water and noble gases 23
    3.3.2 Determination of the water mass 23
    3.3.3 Noble gas analysis..26
  3.4 Results and Discussion 27
    3.4.1 Separation of air and water inclusions...30
    3.4.2 Noble gas temperatures 30
    3.4.3 Noble gas components 31
  3.5 Conclusions and Outlook 35
  3.6 Annex 1..37

    3.6.1 Supplementary Tables 37
    3.6.2 Supplementary Figures 39
  3.7 Annex 2...42
    3.7.1 Noble gas analysis in zebra mussel shells 42
    3.7.2 Noble gas analysis in opalinus clay samples................................42

4 Determination of Holocene cave temperatures from Kr and Xe concentrations in stalagmite fluid inclusions **45**
  4.1 Introduction 45
  4.2 Materials and Methods..47
    4.2.1 Samples..47
    4.2.2 Noble gas analysis...47
  4.3 Results and Discussion 48
    4.3.1 Noble gas temperatures (NGTs)...48
    4.3.2 Suitability of stalagmites for NGT determination 53
  4.4 Conclusions and Outlook 54

5 Application of the noble gas thermometer to fluid inclusions in two Holocene stalagmites from Socotra Island (Yemen) **57**
  5.1 Introduction 57
  5.2 Climate on Socotra Island 59
  5.3 Material and Methods..60
  5.4 Results and Discussion 64
    5.4.1 Noble gas temperatures 64
    5.4.2 Water content and "excess air"...67
  5.5 Conclusions and Outlook 69

6 Trace gas analysis in air inclusions of stalagmites................................**71**
  6.1 Introduction 71
  6.2 Material and Methods..72
    6.2.1 Samples..72
    6.2.2 Gas analysis 73
  6.3 Results and Discussion 73
    6.3.1 Major elemental composition 73
    6.3.2 Stable isotope composition of CO_2 and CH_4 75
  6.4 Conclusions and Outlook 80

7 Synthesis and Outlook...**81**

References **85**

Summary

This work focuses on the determination of noble gas temperatures (NGTs) in water inclusions in stalagmites with the aim to infer the cave temperature prevailing at the time the inclusions were formed. The method is based on the temperature-dependent solubilities of noble gases (He, Ne, Ar, Kr and Xe) in water, which allows calculating the water temperature from measured noble gas concentrations if the salinity and the atmospheric pressure during gas exchange are known. In recent years, stalagmites are increasingly being used as climate archives, as they provide highly resolved and well-dated records of past climate conditions over glacial-interglacial time intervals. Stalagmites contain 0.01-0.1 wt.% of water, which is incorporated into the calcite crystals in the form of water inclusions during stalagmite growth. Noble gas analysis in water inclusions in stalagmites hence offers the potential to directly and quantitatively determine cave temperatures, which is an important prerequisite for the interpretation of $\delta^{18}O$ and $\delta^{13}C$ records in the stalagmite calcite.

To investigate the feasibility of NGT determination in stalagmites, a suitable analytical method to determine noble gas concentrations in stalagmites first had to be developed. Most important for this method is the separation of noble gases released from air inclusions (2-3 vol.%) from those liberated from water inclusions, as only noble gases in water inclusions carry information about the cave temperature. To this end, stalagmite samples are pre-crushed into grains of a defined diameter so that air inclusions are predominantly opened during pre-crushing while water inclusions are largely left intact. This is possible, as air inclusions are usually larger than water inclusions and are more often found along crystal boundaries. Noble gases and water are then extracted from the remaining air and water inclusions in the pre-crushed sample by heating the sample in vacuum to 300-400°C. The small amounts of water liberated from the stalagmite samples (typically 0.5 – 3 mg) are determined manometrically, i.e. by measuring the water vapour pressure in a known volume and at a constant temperature of 40°C. Noble gas concentrations were analysed with overall analytical errors of ca. 2-3% in samples from 11 stalagmites from 8 caves located in different parts of the world (Switzerland, Germany, Turkey, Oman, Yemen).

The interpretation of the noble gas concentrations in terms of a cave temperature turned out to be more difficult than expected, as additional noble gas components were found to be present in stalagmites besides the noble gases released from air and water inclusions. As a result, the concentrations of He, Ne and Ar, which are affected by additional noble gas components (lattice-trapped He and Ne and adsorbed Ar), were excluded from NGT determination in this thesis. Kr and Xe concentrations, however, originate solely from air-saturated water and atmospheric air in a large fraction of the samples analysed. In these samples Kr and Xe concentrations could hence be directly converted into cave temperatures. Calculated NGTs in Holocene samples (1 to 6 ka BP) in stalagmites from regions with different mean annual air temperatures agreed well with modern cave temperatures. This shows that the application of the "noble gas thermometer" to stalagmites is feasible and that stalagmites are a suitable archive for NGT determination. However, we also observed a large variation in NGTs in samples from two Holocene stalagmites from Socotra Island (Yemen), which we attributed mainly to changing partial pressures of Kr and Xe during the gas exchange process due to the accumulation of CO_2 in the air layer around the stalagmite.

The results of this thesis give strong indication that the arrangement of calcite crystals within a stalagmite and the origin of its fluid inclusions (primary or pseudo-secondary) are a crucial prerequisite for the suitability of a stalagmite for NGT determination. Sample selection hence seems to be a key aspect in climate studies using NGTs in stalagmites in the future. In this study, NGTs were more successfully determined in fast growing stalagmites with small and irregularly arranged calcite crystals containing mainly water inclusions of primary origin.

In summary, this study represents an important step towards a better understanding of the geochemical origin of noble gases in stalagmites and their interpretation in terms of cave temperatures. The results of this thesis hence set the basis for a broader application of the "noble gas thermometer" to stalagmites in order to study past climate conditions and their evolution in different regions of the world.

Zusammenfassung

Die Löslichkeit von atmosphärischen Edelgasen (He, Ne, Ar, Kr und Xe) in Wasser ist neben dem atmosphärischen Druck und der Salinität des Wassers auch von der Temperatur des Wassers abhängig. Deshalb kann aus den Edelgaskonzentrationen, die in einer Wasserprobe gemessen werden, die Temperatur des Wassers zum Zeitpunkt des letzten Gasaustausches berechnet werden. Das Ziel dieser Arbeit ist es, diese Methode in Wassereinschlüssen in Stalagmiten anzuwenden, um die so genannte „Edelgastemperatur" des Wassers zu bestimmen, die zum Zeitpunkt des Einschlusses in die wachsenden Kalzitkristalle vorherrschte.

Stalagmiten wurden in den letzten Jahren immer häufiger als Klimaarchive verwendet, da sie über lange Zeiträume wachsen (bis zu mehrere 10^5 Jahre) und gut datierbare, hoch aufgelöste Klimainformationen enthalten, vor allem in der isotopischen Zusammensetzung des Kalzits ($\delta^{18}O$, $\delta^{13}C$). Stalagmiten beinhalten ca. 0.01-0.1 gew.% an Wasser in Wassereinschlüssen, die während des Wachstums des Stalagmiten in die Kalzitkristalle eingeschlossen wurden. Die temperaturabhängige Löslichkeit von Edelgasen in Wasser erlaubt es im Prinzip, die Höhlentemperatur über Edelgasanalysen in den Wassereinschlüssen direkt und quantitativ zu bestimmen. Dies wäre für die Interpretation der $\delta^{18}O$- und $\delta^{13}C$ Zeitreihen in Stalagmiten wichtig, da die isotopische Zusammensetzung des Kalzits unter anderem auch von der Höhlentemperatur abhängt.

Das erste Ziel dieser Arbeit war es, eine Methode zu entwickeln, um Edelgaskonzentrationen in Stalagmiten überhaupt messen zu können. Weil nur Edelgase aus Wassereinschlüssen eine Information über die Temperatur bei der Bildung der Einschlüsse enthalten, ist es wichtig, Edelgase aus Lufteinschlüssen, die auch in Stalagmiten enthalten sind, von Edelgasen aus Wassereinschlüssen zu trennen. Deshalb wurden die Stalagmitenproben zuerst mit einem Mörser auf eine bestimmte Korngrösse zerkleinert. Die Korngrösse beim Zerkleinern wurde so gewählt, dass hierbei vor allem Lufteinschlüsse geöffnet werden und Wassereinschlüsse so weit als möglich intakt bleiben. Dies ist möglich, da Lufteinschlüsse in der Regel grösser sind als Wassereinschlüsse und oft auch entlang von Korngrenzen zu finden sind. Nach dem Zerkleinern der Probe wurden die Edelgase dann im Vakuum in einem 1-stündigen Heizschritt bei 300-400°C aus der Probe extrahiert. Die extrahierte Wassermenge wurde mit einer Druckmessung des Wasserdampfes in einem kalibrierten Volumen und bei einer konstanten Temperatur bestimmt. Damit konnten in dieser Arbeit Edelgaskonzentrationen mit einer Genauigkeit von 2-3% bestimmt werden, in Proben aus gesamthaft 11 Stalagmiten aus 8 verschiedenen Höhlen mit unterschiedlichen Höhlentemperaturen.

Die Interpretation der Edelgaskonzentrationen als Edelgastemperaturen hat sich dann als schwieriger als erwartet herausgestellt. Insbesondere wurden neben den erwarteten Edelgaskomponenten aus den Luft- und Wassereinschlüssen, d.h. atmosphärische Edelgase und Edelgase aus luftgesättigtem Wasser, auch weitere Edelgase aus den Proben extrahiert. Diese Edelgase stammen aus neu identifizierten Edelgaskomponenten in Stalagmiten. Einerseits aus einer Gitterkomponente, die an leichten Edelgasen He und Ne angereichert ist. Andererseits wurde eine zusätzliche Argon-Komponente identifiziert, die durch Adsorption während des Zerkleinerns der Probe zustande kommt. Aus diesem Grund wurden in dieser Arbeit Edelgastemperaturen nur aus den Konzentrationen von Kr und Xe berechnet. In einem grossen Teil der Proben liessen sich Kr und Xe Konzentrationen als einfache Mischung aus atmosphärischer Luft und luftgesättigtem Wasser auffassen, was letztlich ermöglichte Edelgastemperaturen aus Kr und Xe Konzentrationen zu berechnen.

Die Edelgastemperaturen, die in holozänen Stalagmitenproben aus Höhlen in der Schweiz, der Türkei und der Sokotra Insel (Yemen) bestimmt wurden, stimmen gut mit der heutigen Höhlentemperatur überein und liegen damit auch nahe der modernen mittleren Jahrestemperatur. Dies zeigt, dass die Methode der Edelgastemperaturen tatsächlich auch auf Wassereinschlüsse in Stalagmiten übertragbar ist. Allerdings wurde in zwei Stalagmiten auch eine für den untersuchten Zeitraum unrealistisch grosse Streuung der Edelgastemperaturen um den Mittelwert beobachtet. Diese Schwankung kann zumindest teilweise mit reduzierten Partialdrücken von Kr und Xe während des Gasaustausches erklärt werden, da CO_2 in der Luftschicht um den wachsenden Stalagmiten stark angereichert zu sein scheint, was letztlich die Partialdrücke der atmosphärischen Edelgase entsprechend vermindert.

Diese Arbeit zeigt weiter auf, dass die Anordnung der Kalzitkristalle innerhalb eines Stalagmiten, sowie der Ursprung der Wassereinschlüsse (primäre oder pseudo-sekundär) massgebend dazu beitragen, ob die gemessenen Edelgaskonzentrationen erfolgreich in Edelgastemperaturen umgerechnet werden können. So konnten Edelgastemperaturen mehrheitlich in schnell wachsenden Stalagmiten mit kleinen und unregelmässig angeordneten Kristallen und vielen primären Wassereinschlüssen bestimmt werden. Eine sehr sorgfältige licht- und elektronenmikroskopische Untersuchung der Stalagmiten ist deshalb in Zukunft unerlässlich für die Auswahl der Stalagmiten für Edelgastemperaturbestimmung.

Gesamthaft ist diese Arbeit ein wichtiger und notwendiger Schritt, um die geochemische Herkunft von Edelgasen in Stalagmiten zu bestimmen. Dies ist eine wichtige Voraussetzung, um die gemessenen Konzentrationen als Edelgastemperaturen zu interpretieren. Die Resultate der Arbeit liefern somit einen Beitrag zur Bestimmung von Edelgastemperaturen in Stalagmiten, so dass in Zukunft Klimabedingungen und ihre Veränderung in verschiedenen kontinentalen Regionen an Edelgasen in Stalagmiten untersucht werden können.

1

Introduction

1.1 Introduction and scope of work

The concentrations of atmospheric noble gases (He, Ne, Ar, Kr and Xe) dissolved in meteoric waters have become a powerful tool to reconstruct past climatic and environmental conditions, i.e. the temperature and the salinity of the water and the atmospheric pressure at the time of the last gas exchange (Kipfer et al., 2002). This method has so far been successfully applied for reconstructing past groundwater temperatures (e.g. Beyerle et al., 1998; Stute et al., 1995a; Stute et al., 1995b, Weyhenmeyer et al., 2001), for studying mixing processes in lakes (e.g. Aeschbach-Hertig et al., 2002; Hofer et al., 2002; Holzner et al., 2008) and for the determination of past climate conditions from pore waters in unconsolidated sediments (Strassmann et al., 2005; Brennwald et al., 2004; 2005).

The aim of this study is to expand environmental noble gas geochemistry to solid samples containing small quantities of water (several milligrams per gram of sample) in fluid inclusions to reconstruct the environmental conditions prevailing at the time the inclusion was formed. Stalagmites constitute such a new study object of climate research, as they represent an excellent climate archive providing information about past climate conditions for a wide range of continental regions and over timescales which other continental archives fail to cover (Henderson, et al., 2006).

The idea of using noble gas concentrations in fluid inclusions as a direct proxy for cave temperatures emerged many years ago in several noble gas laboratories, but was long hampered by analytical difficulties (pers. comm. B. Marty, CNRS (France), 2005; H. Vonhof, University of Amsterdam (Netherlands), 2006). The first target of this thesis was therefore to develop new methodological concepts to allow the accurate determination of noble gas concentrations in fluid inclusions in stalagmites. In principle it should then be possible to use the measured noble gas concentrations to determine the noble gas temperature (NGT) of the water at the time the inclusions were formed. If this can be analytically and conceptually realized, NGTs will be an extremely valuable new proxy for cave temperatures and a helpful constraint for a more quantitative interpretation of $\delta^{18}O$ and $\delta^{13}C$ records in stalagmites. Moreover, NGTs from fluid inclusions in stalagmites could be used to reconstruct palaeotemperatures on a local and regional scale, as the temperature of cave drip waters reflects the temperature of the overlying soil, and hence the mean annual air temperature outside the cave (McDermott et al., 2005). The structure of the thesis is described in the following section.

1.2 Outline

Scientific background (Chapter 2): This chapter summarizes the conceptual aspects of environmental noble gas geochemistry. It compiles information about atmospheric noble gas concentrations in water and describes the concept of noble gas temperatures (NGTs). It also provides some background information about stalagmites and their use as climate archives.

Experimental methods (Chapter 3): The determination of noble gas concentrations in stalagmites is challenging because of i) the small amounts of gas to be measured, ii) the difficulty of determining the mass of the small amounts of water, which are expected to be liberated from the stalagmites and iii) the presence of air inclusions, which contain much higher concentrations of noble gases than water inclusions, thus masking the temperature dependent noble gas signal in the water inclusions. Chapter 3 describes the analytical protocol for noble gas analysis as well as two extraction techniques for the efficient separation of noble gases released from air and water inclusions. This chapter has been published in Chemical Geology (Scheidegger et al., 2010).

Determination of Holocene cave temperatures from Kr and Xe concentrations in stalagmite fluid inclusions (Chapter 4): In a first study, the analytical protocol developed was applied to stalagmite samples from caves in Switzerland, Turkey and Yemen with mean annual temperatures of between 8°C and 27°C. The results reveal the presence of a clear temperature signal in Kr and Xe concentrations and show also that NGTs calculated from the concentrations of these two noble gases agree well with modern cave temperatures. In addition, the study introduces new conceptual aspects concerning the suitability of stalagmites for NGT determination based on examinations by light and electron microscopy. This chapter has been submitted for publication in Chemical Geology and is currently under review.

Application of the noble gas thermometer to fluid inclusions in two Holocene stalagmites from Socotra Island (Yemen) (Chapter 5): In a second study, noble gas concentrations were analysed in ca. 40 samples from two Holocene stalagmites from Socotra Island with the aim to determine NGTs over the time interval covered by the samples (the last 4 and 10 ka). The results revealed a large variation of NGTs, which is very unlikely to be associated with actual cave temperature variations over the Holocene. The observed NGT variation is explained by varying conditions during the gas exchange process of noble gases between the cave air and the fluid inclusion water, which seem to prevent an accurate NGT determination. The chapter also introduces the water content and the amount of "excess air" in the samples analysed as new potential climate proxies in stalagmites. This chapter is in preparation for publication.

Trace gas analysis in air inclusions (Chapter 6): Apart from dissolved atmospheric noble gas concentrations in water inclusions, trace gases in air inclusions in stalagmites could potentially be used as a proxy for the past composition of the cave air. During this thesis, first attempts to analyse the major atmospheric gases as well as their isotopic composition were undertaken to assess the potential of this new climate proxy in stalagmites. The semi-quantitative and preliminary results are presented in chapter 6.

Conclusions and Outlook (Chapter 7): The last chapter summarizes the most important findings of this thesis and compiles a selection of future research tasks.

2

Scientific background

2.1 Noble gases

Noble gases - helium (He), neon (Ne), argon (Ar), krypton (Kr) and xenon (Xe) - represent the group Nr. 8 in the periodic table of the elements, which is characterized by elements with a "full" outer electron shell. This gives noble gases little tendency to participate in biogeochemical reactions. Noble gases hence only take part in physical reactions (e.g. solution, adsorption, diffusion), which makes them ideal trace gases to study physical processes. A summary of physical characteristics of atmospheric noble gases important for this thesis is given in Table 2.1.

2.1.1 Noble gases in water as environmental tracers

An important physical process, which allows studying environmental water systems using noble gas concentrations, is the dissolution of noble gases in water by gas exchange (for a review see Kipfer et al., 2002). Noble gases enter the water from the atmosphere, which is the major reservoir of terrestrial noble gases. The process of air-water partitioning is described by Henry's law:

$$p_i = H_i(T,S) \cdot C_i^* \qquad i = He, Ne, Ar, Kr, Xe$$

Henry's law states that the equilibrium concentration of a noble gas C_i^* in water depends on its partial pressure p_i in the gas phase and on the Henry coefficient, which in turn depends on the temperature (T) and the salinity (S) of the water.

Table 2.1: Characteristic properties of the noble gases He, Ne, Ar, Kr and Xe. Given are the volume fraction of the element in dry air (v_i), the relative abundance (R_i) of each isotope in air (Ozima and Podosek, 2002) and the atomic diameter (d_i) (Huber et al., 2006).

Noble gas	v_i (-)	d_i (Å)	Isotope	R_i (%)
He	$5.24 \cdot 10^{-6}$	2.55	^{3}He	0.00014
			^{4}He	~ 100
Ne	$1.81 \cdot 10^{-5}$	2.82	^{20}Ne	90.5
			^{21}Ne	0.268
Ar	$9.34 \cdot 10^{-6}$	3.45	^{36}Ar	0.3364
			^{38}Ar	0.0632
			^{40}Ar	99.6
Kr	$1.14 \cdot 10^{-6}$	3.65	^{78}Kr	0.347
			^{80}Kr	2.257
			^{82}Kr	11.523
			^{83}Kr	11.477
			^{84}Kr	57.00
			^{86}Kr	17.398
Xe	$8.7 \cdot 10^{-6}$	4.04	^{124}Xe	0.0951
			^{126}Xe	0.0887
			^{128}Xe	1.919
			^{129}Xe	26.44
			^{130}Xe	4.070
			^{131}Xe	21.22
			^{132}Xe	26.89
			^{134}Xe	10.430
			^{136}Xe	8.857

Figure 2.1 and Table 2.2 show that the solubility of noble gases in water decreases as the temperature increases. Also, the temperature dependence of the solubility is stronger for the heavy noble gases Ar, Kr and Xe than for the light noble gases He and Ne. A temperature change, for instance, from 0°C to 30°C reduces the solubility of He by 11%, of Ne by 24% of Ar by 47%, of Kr by 55% and of Xe by 62%.

Table 2.2: Equilibrium concentrations of He, Ne, Ar, Kr and Xe in cm^{3}STP/g at sea level for different water temperatures T and salinities S (1‰=10g/kg) using the Henry coefficients recommended by Kipfer et al. (2002).

T	10°C		20°C		30°C	
S	0‰	10‰	0‰	10‰	0‰	10‰
He (10^{-8})	4.64	4.39	4.48	4.24	4.36	4.15
Ne (10^{-7})	2.02	1.89	1.85	1.75	1.72	1.63
Ar (10^{-4})	3.86	3.59	3.12	2.92	2.60	2.44
Kr (10^{-8})	9.10	8.45	6.96	6.49	5.52	5.18
Xe (10^{-8})	1.32	1.22	0.95	0.88	0.72	0.67

The concentrations of dissolved noble gases are usually denoted as gas volumes (cm^3) at standard conditions (STP: T = 0°C, p = 1013 mbar) per unit water mass (g); i.e. cm^{3}STP/g. In this work, the noble gas solubilities are calculated from the empirical parameterizations of Weiss (1971) for He and Ne, Weiss (1970) for Ar, Weiss and Kyser (1978) for Kr and Clever (1979) for Xe, as recommended by Kipfer et al. (2002).

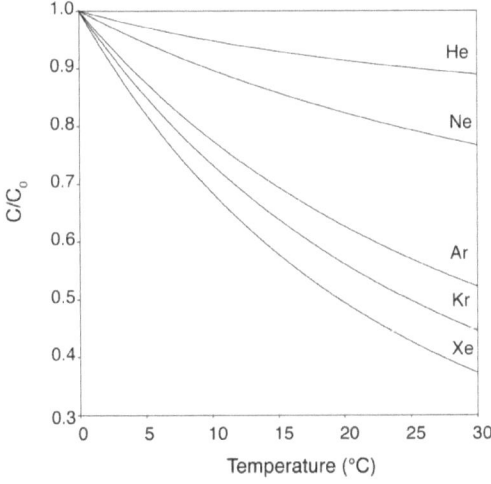

Figure 2.1: Temperature dependence of equilibrium concentrations of noble gases in water, calculated using the Henry coefficients recommended by Kipfer et al. (2002) and normalized to the concentrations at 0°C.

2.1.2 Determination of "noble gas temperatures (NGTs)"

Based on the physical parameters describing Henry's law, the concentrations of He, Ne, Ar, Kr and Xe dissolved in water depend on the atmospheric pressure p, the temperature T and the salinity S of the water at the time of the last gas exchange with the atmosphere. This offers the possibility to determine p, T and S prevailing during the last gas exchange from dissolved atmospheric noble gas concentrations measured in any meteoric water sample, which has been disconnected from the atmosphere (Kipfer et al., 2002). For instance, noble gas concentrations can be used to determine the water temperature (the noble gas temperature NGT) by least squares fitting of measured noble gas concentrations (Aeschbach-Hertig et al., 1999) if the salinity of the water and the atmospheric pressure during gas exchange are known. This principle was successfully applied to groundwater samples to reconstruct NGTs during the last glacial period and to determine the temperature shift at the transition from the LGM to the Holocene (e.g. Kreuzer et al., 2009; Ma et al., 2004; Stute et al., 1992; Stute et al., 1995b; Weyenmeyer et al., 2000).

The least squares fitting models developed by Aeschbach-Hertig et al. (1999; 2000) include the variable "excess air", which describes an atmospheric noble gas component originating from the total or partial dissolution of air bubbles in water, so that the totally measured noble gas concentrations in water samples are usually composed of noble gases from air and from air-saturated water (for details about "excess air" see Aeschbach-Hertig et al., 2000; Heaton and Vogel, 1981; Klump et al., 2007; Klump et al., 2008). The least squares fitting models can hence only be used to interpret binary mixtures of noble gases from air-saturated water (ASW) and atmospheric air. If additional noble gas components are present, for instance radiogenic ^{4}He, which is often found in groundwater samples, their concentrations must be excluded from the determination of NGTs. The number of variables P to be reconstructed (in this thesis the temperature T and the amount of "excess air") must not exceed the number of noble-gas concentrations N reflecting these variables. If P < N, the variables are estimated by least squares fitting according to Aeschbach-Hertig et al. (1999). In any case, the concentrations of at least two noble gases are needed to determine the two parameters T and "excess air". In this case the number of unknown variables P equals the number of noble gas concentrations N and the two variables are calculated by algebraically solving the set of two simultaneous equations.

2.2 Stalagmites

Stalagmites are secondary deposits of calcium carbonate ($CaCO_3$), which precipitate in the cave environment from a supersaturated solution covering the surface of the stalagmite. In the soil zone above the cave, CO_2 levels are elevated by a factor of 10-100 ($v_{CO2,soil}$ = 0.1-3.5 vol.%) with respect to the atmosphere ($v_{CO2,atm}$ = 0.036 vol.%) due to microbial activity and plant respiration. Meteoric waters percolating through the soil zone interact with the CO_2 and hence dissolve the limestone bedrock. In the cave atmosphere CO_2 levels are again lower than in the soil zone ($v_{CO2,cave}$ = 0.06-0.6 vol.%), which leads to CO_2 degassing from the drip water to reach a new equilibrium. Subsequently the water is supersaturated in carbonate and stalagmites are deposited according to the chemical reaction $Ca^{2+} + 2HCO_3^- = CaCO_3 + CO_2 + H_2O$ (McDermott et al., 2005 and Figure 2.2). Stalagmite growth rates range from 20-1000 µm/a and mainly depend on the cave temperature, on the drip rate, the calcium saturation of drip waters and on the CO_2 partial pressure difference between soil and cave air (McDermott et al., 2005).

2.2.1 Stalagmites as climate archives

In the last decades, the interest in stalagmites as climate archives has increased strongly among palaeoclimatologists as stalagmites grow continuously over long time intervals ($10^3 - 10^5$ years) and can be precisely dated by U/Th-series techniques (Ivanovich and Harmon, 1993). Most other climate archives covering similar timescales, e.g. ice cores and ocean sediments, are limited to regions of permanent ice cover and marine regions. Stalagmites are found in continental regions in almost all parts of the world and hence offer the possibility to study past climate conditions in regions where no other climate archives are available (McDermott et al., 2005, Henderson, 2006).

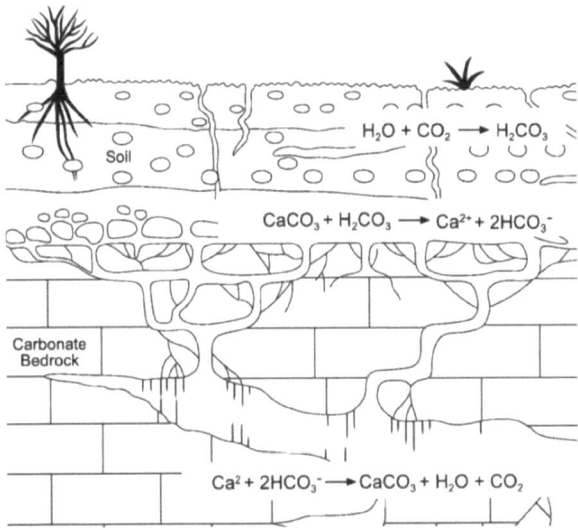

Figure 2.2: Illustration of a cave and the processes leading to the deposition of stalagmites (adapted from Fairchild et al., 2006).

The stable isotope composition of the calcium carbonate ($\delta^{13}C_{calcite}$ and $\delta^{18}O_{calcite}$) is the most widely used climate proxy in stalagmites. It is used to infer e.g. the timing of major climate events in the past (Burns et al., 2001; Fleitmann et al., 2009; Genty et al., 2003; Liu et al., 2010; Spötl and Mangini, 2002), to reconstruct changes in precipitation (Burns et al., 2003; Fleitmann et al., 2003a, Fleitmann et al., 2007, Fleitmann et al., 2009; Griffith et al., 2010b) and to determine the type of plant cover above the cave (Cosford et al., 2009; Dorale et al., 1998; Vaks et al., 2010). Other proxies in stalagmites include for instance trace elements to reconstruct palaeohydrological conditions (Buhl et al., 2007; Fairchild et al., 2009; Griffiths et al., 2010a; Treble et al., 2005; Wynn et al., 2010;), plant derived biomarkers to reconstruct land use changes (Blyth et al., 2007; Blyth and Watson, 2009) and organic material that provides information about the vegetation regime and bacterial activity in the soil above the cave (Perett et al., 2005; Blyth et al., 2010).

Apart from this proxy information, that is stored in the stalagmite calcite, stalagmites contain fluid inclusions, which are incorporated into the calcite crystals during stalagmite growth (Schwarcz and Harmon, 1976; Kendall and Broughton 1978). The formation of fluid inclusions along imperfections and surface irregularities of growing crystals is a common process in minerals, which precipitate from a supersaturated solution (Goldstein and Reynolds 1994). According to Roedder (1984), discontinuities such as e.g. fast changing growth rates, changes in the saturation of the solution and rapid changes in growth conditions seem to foster the formation of fluid inclusions. In stalagmites, the analysis of the δD signature of the fluid inclusion water has been used to reconstruct palaeotemperatures during the early Holocene (Zhang et al., 2008), during the last glacial period (Harmon, 1979) and to study the variability of precipitation during the ultimate and the penultimate glacial periods (Fleitmann et al., 2003b; McGarry et al., 2004; Schwarcz and Harmon, 1976).

Whereas $\delta^{13}C_{calcite}$ and $\delta^{18}O_{calcite}$ records are now widely available over glacial-interglacial timescales and for many different geographical regions, the interpretation of these records is still challenging. This is because the stable oxygen and carbon isotope signature of the calcite depends on many different processes, e.g. on the cave temperature, on the amount and the source of precipitation and, over glacial-interglacial timescales, also on the amount of the global ice volume (see Lachniet et al., 2009; Darling et al., 2005). A direct cave temperature proxy in stalagmites hence offers the possibility to disentangle the various effects influencing the stable isotope composition of the calcite to a certain extent. Also, it will eventually allow to quantitatively determining the isotope composition of the drip water, using the empirical equations for the temperature dependent isotope fractionation between calcite and water (e.g. Craig, 1965; Kim and O'Neil, 1997). In addition, cave temperatures commonly remain constant throughout the year and hence provide good estimates of the annual mean ambient air temperature outside the cave (Poulson and White, 1969; Smithson, 1991; McDermott, 2005). Hence, direct reconstruction of cave temperatures would allow to determine palaeotemperatures on a local scale, over long timescales and for a wide range of regions like e.g. continental, tropical and high altitude regions.

2.2.2 Fluid inclusions in stalagmites

Microscopic investigations carried out within this thesis showed that two distinct types of fluid inclusions are present in stalagmites, i.e. air and water inclusions. The two types of inclusions are clearly distinguishable under the microscope, as both types differ by their optical appearance and their arrangement within the stalagmite. Water inclusions account for 0.01 – 0.1% of the stalagmite weight (Kendall and Broughton, 1978; Schwarcz and Harmon, 1976) and air inclusions for up to 3% of the stalagmite volume (Scheidegger et al., 2007; Badertscher, 2007). While air and water inclusions are present in all stalagmites, their abundance, their size and their shape strongly vary between stalagmites. A selection of fluid inclusions found in stalagmites investigated within this thesis is shown Figure 2.3.

2.2.3 Noble gas concentrations in fluid inclusions

In recent years, much effort has been put into the development of new methods that allow direct determination of cave temperatures in stalagmites, e.g. the liquid-vapour homogenisation temperature in fluid inclusions (Kruger et al., 2007, 2008). Also, the temperature dependent equilibrium fractionation of oxygen isotopes between fluid-inclusion water and calcite has been used to reconstruct Holocene temperature variations in Peru and Indonesia (Griffiths et al., 2010; van Breukelen et al., 2008) and to determine the temperature difference between the termination II and the Eemian period (Wainer et al., 2010). Another method for direct cave temperature reconstruction is 'clumped isotope' thermometry, which has been used to reconstruct glacial-interglacial temperature variations in Israel (Affek et al., 2008). The latter two methods are related to the isotope composition of the calcite and require equilibrium conditions during calcite precipitation to yield correct cave temperatures. Hence, the application of these methods is so far limited to stalagmites that are deposited in isotope equilibrium.

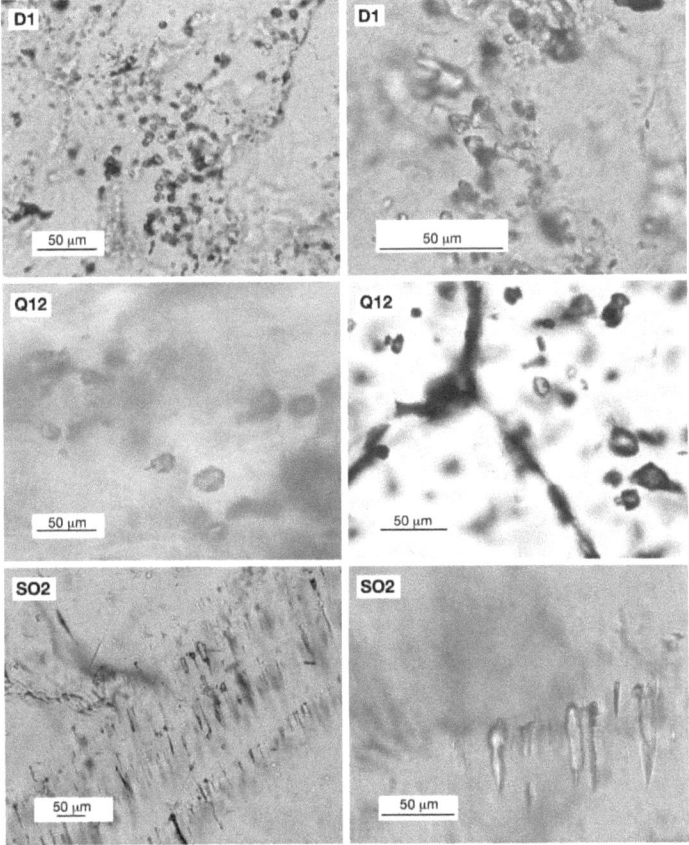

Figure 2.3: Photomicrographs of thin sections showing fluid inclusions in three stalagmites (further information on the stalagmites is given in section 2.5). Stalagmite D1 has a high abundance of water inclusions, which are arranged in bands parallel to the surface of the growing stalagmite. In stalagmite Q12 water inclusions are sparse and often contain a gas bubble. Stalagmite SO2 consists of several cm long columnar crystals with strongly elongated air and water inclusions, which may have been formed by crystal coalescence (Kendall and Broughton, 1976).

In contrast, noble gas concentrations in stalagmite fluid inclusions are independent of the isotope composition of the calcite and governed by the well constraint and "simple" physics of gas exchange between the fluid inclusion water and the surrounding air. Atmospheric noble gases are dissolved in the drip water and the water film covering the growing stalagmite according to Henry's Law (Figure 2.4). Hence, the noble gas concentrations in the water depend on the atmospheric pressure p, the salinity S and temperature T of the water. Incorporation of the equilibrated drip water in fluid inclusions separates the drip water from the cave atmosphere and hence prevents further gas exchange. As a result, the noble gas concentrations in the water-filled inclusions provide in principle direct information on the temperature in the cave at the time of fluid formation.

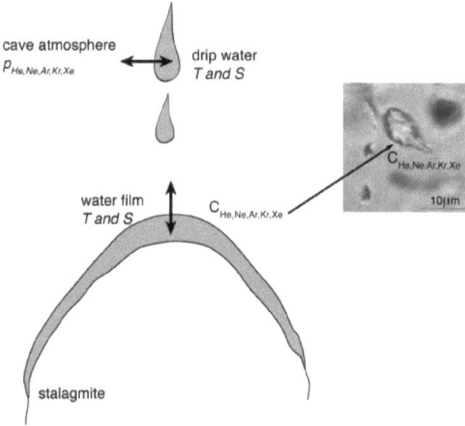

Figure 2.4: Schematic illustration of the gas exchange processes in the drip water and in the water film on the growing stalagmite.

We expect the noble gas concentrations in the thin film of drip water on the stalagmite surface to be in solubility equilibrium with the surrounding air, because the equilibration time t (t ≈ $d^2/2D$) is ~1 min, assuming a water film thickness d of ~ 0.1 mm (Dreybrodt, 1980) and a molecular diffusivity D of noble gases in water of ~10^{-9} to 10^{-10} m^2/s (Ozima and Podosek, 2002). Also, the inventory of noble gases in the atmosphere can safely be assumed to remain constant over the timescales (1-100 ka) relevant for this thesis. The relative abundance of noble gases in cave air can be expected to be equal to their relative abundance in the atmosphere, as there are no significant noble gas sources or sinks in caves. Mass spectronomic analyses of modern air samples from Sofular cave (Turkey) indeed showed no deviations in their noble gas composition to atmospheric air. However, the absolute partial pressures of noble gases may be different than in the free atmosphere, due to the accumulation of e.g. CO_2 produced during calcite precipitation. The salinity of groundwater in karst systems is usually low enough to allow the assumption of S = 0 g/kg and the atmospheric pressure is determined by the altitude of the cave and can be calculated using the barometric formula.

As a result of these considerations, noble gas concentrations in stalagmites are expected to be related only to the temperature of the drip water and to the amount of noble gases released from air inclusions, denoted as "excess air" A in cm^3 air per g of released water. This allows - in principle - a direct determination of NGTs in stalagmites.

2.3 Analysis of stable isotopes

This thesis primarily focuses on the analysis of noble gases in water inclusions with the goal to determine NGTs. However, as mentioned in section 2.2, stalagmites also contain air inclusions, which may be used as a climate proxy similar to air trapped in ice sheets (e.g. Petit et al., 1999; Jouzel et al., 2007; Andersen et al., 2004). In collaboration with the Institute of Physics in Bern we conducted a pilot study on the analysis of trace gases in air inclusions and determined the abundance of major atmospheric gases as well as their isotope composition to assess the potential of air inclusions as a new proxy in stalagmites. This section compiles some background information about stable isotopes of carbon and oxygen, which are the main elements in the gases we analysed in air inclusions (O_2, CO_2, CH_4).

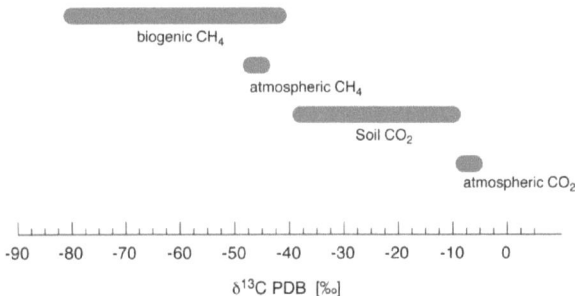

Figure 2.5: Typical values of $\delta^{13}C$ for selected natural compounds, which are relevant for the interpretation of stable isotope analysis in air inclusions in stalagmites (Faure and Mensing, 2004; Darling et al., 2005).

Isotopes are atoms of the same element, which vary in their mass due to different numbers of neutrons in the core. Physical, chemical and biological processes usually show a small dependence on the atomic mass of the reacting elements, so that some isotopes of one element are more reactive than others. This leads to mass fractionation and hence to specific fractionated isotope compositions of the produced compounds. The fractionation of isotopes of the same element can be determined experimentally and is indicated by the fractionation factor $\alpha = R_a/R_b$ where R is the ratio of the heavy to the light isotope and a and b are compounds or phases. The isotope composition of a compound is then usually denoted as the ‰ - deviation from a standard sample (marine carbonate PDB for $\delta^{13}C$ and standard mean ocean water SMOW for $\delta^{18}O$). Positive values indicate that the element is enriched in the heavier isotope relative to the composition of the standard, whereas negative values stand for depletion in the heavy isotope. Typical carbon isotope compositions of natural compounds relevant for this thesis are listed in Figure 2.5.

2.4 Sample description

In this thesis stalagmite samples from caves in Oman, Yemen (Socotra Island), Turkey, Germany and Switzerland were analysed (see Figure 2.6). The caves are located in different climatic zones, i.e. in temperate, in Mediterranean and in monsoonal climates and hence cover a wide range of temperatures usually observed in meteoric water systems. The caves and stalagmites are briefly described in the following section and photographs (Figures 2.7 and 2.8) are shown for those stalagmites, which were studied in more detail within this thesis (D1, SO2, V1 and P3).

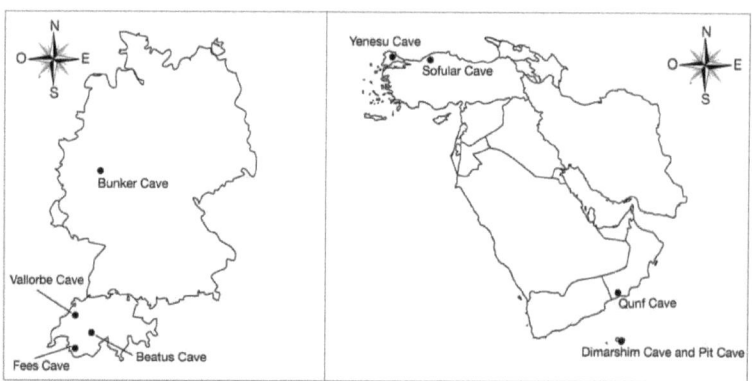

Figure 2.6: Two maps showing the location of the caves, where stalagmites were collected for this thesis (left side: Middle Europe, right side: Middle East).

Beatus Cave
Switzerland, 46°41'N 7°48'E, stalagmite F and G
Beatus cave (870 m.a.s.l.) is located in central Switzerland and has a modern cave temperature of 9°C. The stalagmite samples we used for noble gas analysis have not been dated yet, but were roughly estimated to have an age of 50 to 100 ka BP (pers. comm. P. Häuselmann, SISKA, La Chaux-de-Fonds, Switzerland, 2008).

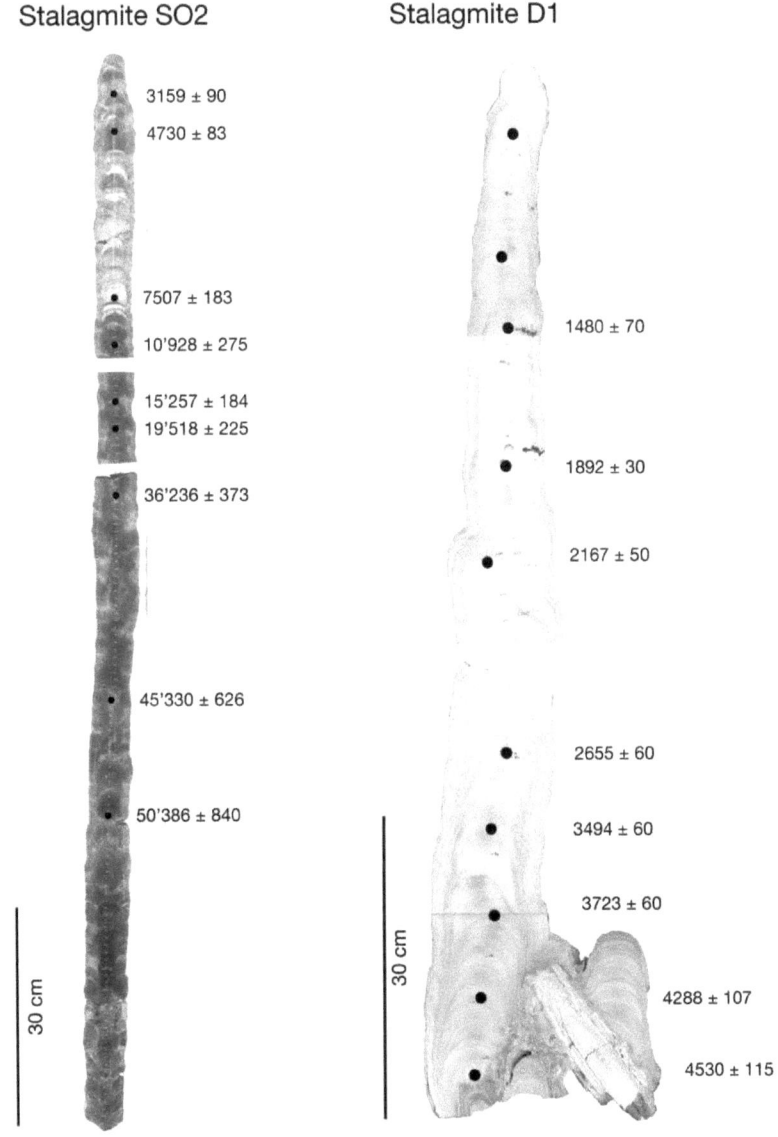

Figure 2.7: Photograph of the stalagmites SO2 and D1. Absolute U/Th-ages are indicated in black in years BP 1950 (Fleitmann et al., 2007, 2009).

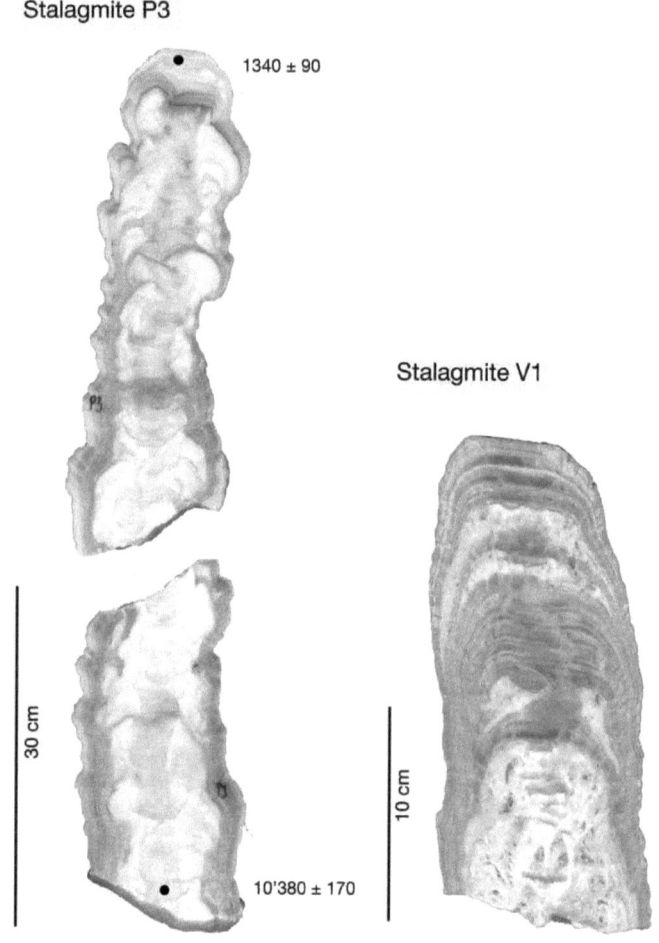

Figure 2.8: Photograph of the stalagmite V1 and P3. Absolute U/Th-ages are indicated in black in years BP 1950.

Bunker Cave
Germany, 51°22' N, 7°40' E, stalagmite BU-U (called stalagmite D in chapter 3)
Bunker cave (180 m.a.s.l.) is situated in the Sauerland in Germany and has a modern cave temperature of 10°C. We determined noble gas concentrations in samples with an age of ~ 7 ka BP from stalagmite BU-U, which was also used by Kluge et al. (2008) for noble gas analysis and NGT determination.

Dimarshim Cave
Socotra Island (Yemen), 12°33'N 53°41'E, stalagmite D1 (called stalagmite A in chapter 3)
Dimarshim Cave is located on Socotra Island in the Northern Indian Ocean. The cave lies at 350 m.a.s.l. and has a modern cave temperature of ~ 27°C. Samples from stalagmite D1 covering the last 4.4 ka BP were used for noble gas analysis. Stalagmite D1 was also analysed for $\delta^{18}O_{calcite}$ at the University of Bern to investigate the evolution of the Asian Monsoon (Fleitmann et al., 2007). Stalagmite D1 consists of an opaque calcite and has clearly visible alterations of white and light-brown laminae (Figure 2.7).

Fees Cave
Switzerland, 46°13'N 7°0'E, stalagmite H
Fees cave is situated at 840 m.a.s.l. in southern Switzerland. The modern cave temperature is 8°C. Samples used for noble gas analysis are from stalagmite H.

Pit Cave
Socotra Island (Yemen), 12°25'N 53°58'E, stalagmite P3
Pit cave is located on a plateau on Socotra Island at 480 m.a.s.l. The modern cave temperature is ~ 28°C. The samples we used for noble gas analysis are from stalagmite P3 and cover the last 10 ka. Stalagmite P3 consists of a white to grey and mostly opaque calcite (Figure 2.8).

Qunf cave
Oman, 17°10'N 54°18'E, stalagmite Q12 (called stalagmite B in Chapter 3)
Qunf cave is located in Southern Oman at 650 m.a.s.l and has a modern cave temperature of 27°C. We analysed noble gas concentrations in samples from stalagmite Q12, which is not dated yet, but was actively growing at the time it was collected in 2002. Stalagmite Q12 is characterized by a partly opaque and partly translucent white calcite. A stalagmite from the same cave (Q5) was used to study the forcing of the Holocene Monsoon in the East Asian region (Fleitmann et al., 2003).

Sofular Cave
Turkey, 41°25'N 31°56'E, stalagmites SO2, SO3, SO4 (SO2 is called stalagmite C in chapter 3)

Sofular cave lies in north-eastern Turkey and has a modern cave temperature of 11.8°C ± 0.1°C. Stalagmites from Sofular cave were used for both noble gas analysis (SO2, SO3, SO4) as well as for trace gas analysis in air inclusions (SO2). Most samples we analysed are of Holocene age with a few older samples, which were deposited during the last ice age. All stalagmites from Sofular cave consist of a translucent light brown to ochre calcite (Figure 2.7). The oxygen isotope composition of stalagmites from Sofular cave was used to infer the timing and the climatic impact of Greenland interstadials (Fleitmann et al., 2009).

Vallorbe Cave
Switzerland, 46°42'N 6°20'E, Stalagmite V1
Vallorbe cave is located 770 m.a.s.l. in the western part of Switzerland. The mean annual temperature in the region is ~ 8°C. Samples from stalagmite V1 are not yet dated and were used for noble gas analyses. Stalagmite V1 is characterized by a partly opaque and partly translucent white to brown calcite (Figure 2.8).

Yenesu Cave
Turkey, 41°37'N 27°57'E, stalagmite Y4
Yenesu cave, situated on the European part of western Turkey at 650 m.a.s.l., has a modern cave temperature of 12°C. Samples from stalagmite Y4, covering the last 1.5 ka were used for trace gas analysis in air inclusions. Stalagmite Y4 is translucent and has a very high abundance of large air inclusions. The stable isotope composition of the calcite has been measured at the University of Bern, but the data has not yet been published.

3

Accurate analysis of noble gas concentrations in small water samples and its application to fluid inclusions in stalagmites

This chapter has been published in Chemical Geology (Scheidegger et al., 2010). The supplementary information given in Annex 1 and Annex 2 at the end of this chapter was not included in the Chemical Geology paper.

Abstract The concentrations of dissolved noble gases in water are widely used as a climate proxy to determine noble gas temperatures (NGTs); i.e., the temperature of the water when gas exchange last occurred. In this paper we make a step forward to apply this principle to fluid inclusions in stalagmites in order to reconstruct the cave temperature prevailing at the time when the inclusion was formed. We present an analytical protocol that allows us accurately to determine noble gas concentrations and isotope ratios in stalagmites, and which includes a precise manometrical determination of the mass of water liberated from fluid inclusions. Most important for NGT determination is to reduce the amount of noble gases liberated from air inclusions, as they mask the temperature-dependent noble gas signal from the water inclusions. We demonstrate that offline pre-crushing in air to subsequently extract noble gases and water from the samples by heating is appropriate to separate gases released from air and water inclusions. Although a large fraction of recent samples analysed by this technique yields NGTs close to present-day cave temperatures, the interpretation of measured noble gas concentrations in terms of NGTs is not yet feasible using the available least squares fitting models. This is because the noble gas concentrations in stalagmites are not only composed of the two components air and air saturated water (ASW), which these models are able to account for. The observed enrichments in heavy noble gases are interpreted as being due to adsorption during sample preparation in air, whereas the excess in He and Ne is interpreted as an additional noble gas component that is bound in voids in the crystallographic structure of the calcite crystals. As a consequence of our study's findings, NGTs will have to be determined in the future using the concentrations of Ar, Kr and Xe only. This needs to be achieved by further optimizing the sample preparation to minimize atmospheric contamination and to further reduce the amount of noble gases released from air inclusions.

3.1 Introduction

The concentrations of atmospheric noble gases (He, Ne, Ar, Kr, Xe) dissolved in water reflect the temperature (T) and salinity (S) of the water and the atmospheric pressure (p) that prevailed during the last gas exchange with the atmosphere, because the solubilities of noble gases are a well-defined function of these parameters (see e.g. Kipfer et al., 2002; Stute and Schlosser, 2000). This approach has been widely used by least-squares fitting of the measured noble gas concentrations (Aeschbach-Hertig et al., 1999; Hall and Ballentine, 1996) to reconstruct past soil temperatures from the concentrations of noble gases in groundwater (Aeschbach-Hertig et al., 2000; Weyhenmeyer at al., 2000; Beyerle et al., 1998; Kipfer et al., 2002; Stute et al., 1995) and to determine past environmental conditions from noble gas concentrations in the pore waters of unconsolidated sediments (Brennwald et al., 2004, 2005; Strassmann et al., 2005).

The same environmental information (T, S and p) may also be preserved in stalagmites, as they contain approximately 0.1 wt.% of water in fluid inclusions. Stalagmites are widely used as climate archives as they provide high-resolution oxygen and carbon isotope records in continental regions covering long timescales (e.g. Fleitmann et al., 2003; McDermott, 2004; McDermott et al., 2005). However, only few proxies exist that allow quantitative reconstruction of cave temperatures, e.g. the analysis of $\delta^{18}O$ in both the fluid inclusion water and the calcite (van Breukelen et al., 2008), the measurement of the liquid-vapour homogenization temperature in fluid inclusions (Krüger et al., 2007) or "clumped isotope thermometry" (Affek et al., 2008). However, these methods suffer from large uncertainties due to archive-specific and analytical shortcomings.

Dissolved atmospheric noble gas concentrations in stalagmite fluid inclusions are directly related to the cave temperature that prevailed during gas-water partitioning according to the physically well-constrained process of gas exchange. Noble gas concentrations can therefore potentially be used to determine the cave temperature (in this paper called noble gas temperature NGT) at the time when the inclusion was formed (see also Kluge et al. 2008). In most caves, temperatures remain constant throughout the year at approximately the annual mean soil temperature. Accordingly, temperature reconstructions from stalagmites provide good estimates of the annual mean temperature outside the cave (McDermott, 2004). In addition, a knowledge of the cave temperature is important in disentangling the various effects that influence the isotopic composition of the calcite, and is thus crucial for the interpretation of oxygen isotope records, e.g. in terms of rainfall amount or of sources of precipitation (Lachniet, 2009).

The determination of noble gas concentrations dissolved in the fluid inclusion water in stalagmites was until recently not feasible, mainly because of the existence of inclusions filled with air and the lack of adequate extraction methods for the separation of air and water inclusions (Ayliffe et al., 1993; Scheidegger et al., 2007b). Noble gases are much more abundant in air than in air-saturated water (ASW) and thus a high abundance of air inclusions leads to large excesses in the measured noble gas concentrations relative to the values expected for solubility equilibrium. Excesses of atmospheric noble gases are also observed in groundwater, where they result from the addition of atmospheric air from partly or totally dissolved air bubbles to the water (Holocher et al., 2003; Kipfer et al., 2002; Klump et al., 2007). Groundwater samples typically contain "excess air" amounts of around 100% ΔNe (100% supersaturation of Ne relative to its concentration in ASW corresponds to ~10^{-2} cm^3 STP of air per cm^3 of water). In contrast, ΔNe values in stalagmites are commonly much larger, i.e. on the order of up to 1000 to 10'000% (Ayliffe et al., 1993; Kluge, 2008; Scheidegger et al., 2007b).

For groundwater samples conceptual models were developed that allow the determination of NGTs even in the presence of moderate amounts of "excess air" (Aeschbach-Hertig et al., 1999). However, for large ΔNe amounts above ~1000% it is not possible to estimate reasonable NGTs from the measured noble gas concentrations. As a consequence, for stalagmite samples the amount of ΔNe has to be reduced by a factor of 10 – 100.

We present an analytical protocol that allows accurate determination of noble gas concentrations in stalagmite samples with a precision of 2-4%. The method includes a way of separating noble gases released from air inclusions from those released from water inclusions to reduce the amount of "excess air". It also includes a precise manometrical determination of water masses of only a few milligrams. The determination of the water mass is a key element of this method, because well-constrained NGTs cannot reliably be derived from noble gas abundance ratios, especially if ΔNe values exceed ~100% (e.g. Kluge et al., 2008).

We will show in section 3.4, that a reliable NGT determination from our stalagmite data is so far not feasible because the available least squares fitting models (Aeschbach-Hertig et al., 1999) are not adequate to conceptually describe the measured noble gas concentrations in stalagmites. We observed "unusual" excesses in light as well as heavy noble gases relative to a simple binary mixture of noble gases from air and ASW. The origin of these noble gas excesses will be explained and discussed in this paper. Nevertheless, a relatively large fraction (40%) of samples analysed by the currently best experimental protocol yield NGTs in agreement with the respective cave temperature, although with large absolute errors.

3.2 Fluid inclusions in stalagmites

During stalagmite growth, minute quantities of drip water and cave air are trapped in the calcite as fluid inclusions that either contain air, water or both (up to 20 vol.% of air, Fig. 3.1). Microscopic analysis of several thin sections of stalagmites from Oman and Yemen (Fleitmann et al., 2007) showed that the two types of fluid inclusions differ clearly in their size and in their position within the calcite (Scheidegger et al., 2007a). Water inclusions (1-50 µm) are always found within calcite crystals (intra-crystalline) and their abundance typically ranges from 0.01 to 0.1 wt.% (Kendall and Broughton, 1978; Schwarcz and Harmon, 1976). Air inclusions are either located within or between calcite crystals (intra-crystalline air inclusions vs. inter-crystalline air inclusions). The intra-crystalline air inclusions are similar in shape to the water inclusions, whereas the inter-crystalline air inclusions have more angular forms and larger sizes (20-80 µm, Scheidegger et al., 2007a).

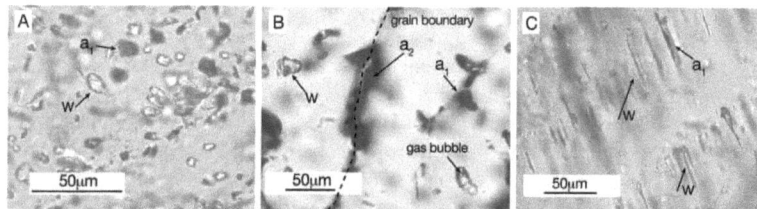

Figure 3.1: Photographs of thin sections from three different stalagmites (A, B and C). Water inclusions (w) and air inclusions (a_1 intra-crystalline, a_2 inter-crystalline) are indicated. The shapes of water inclusions range from elliptical (in A) to strongly elongated (in C) and sometimes contain a gas bubble (in B). In contrast to stalagmite B, where water inclusions are randomly distributed, stalagmites A and C show water inclusions as laminae aligned parallel to the growing surface of the stalagmite.

We did not observe air inclusions connected to the surface of the stalagmite; i.e., the air inclusions most likely do not exchange gases with air in the cave after their formation. Therefore, air inclusions potentially contain information about the composition of the cave atmosphere and about the biological activity in the soil above the cave at the time when the air was trapped in the growing stalagmite (Badertscher et al., 2007). Estimated from microscopic photographs, the air content of a stalagmite is 2-3 vol.%. Inter-crystalline air inclusions account for more than 90% of this (Scheidegger et al., 2007a).

3.3 Methods

In our analytical protocol three subsequent steps finally lead to the determination of the noble gas concentrations: 1) extraction of water and noble gases from fluid inclusions (section 3.3.1); 2) determination of the mass of the extracted water (section 3.3.2); and 3) mass spectrometric measurement of noble gas amounts (section 3.3.3).

3.3.1 Extraction of water and noble gases

Extraction method I
Since inter-crystalline air inclusions account for most of the total air content of a stalagmite, separation of a stalagmite sample into individual crystals significantly reduces the abundance of air inclusions. Optical examinations of different size fractions of crushed calcite grains from a stalagmite (stalagmite A in Fig. 3.1) showed that most grains smaller than 300 µm consist of single crystals, whereas larger grains are aggregates of several crystals. In extraction method I, stalagmite samples of about 3 g were therefore pre-crushed in a mortar into grains ≤ 300 µm in diameter by striking the pestle with a hammer. After each hammer stroke the crushed material was sieved and grains > 300 µm in diameter were separated from the rest of the sample. The procedure was repeated until all grains were ≤ 300 µm in diameter. Twelve samples were pre-crushed in air, and 6 samples in a glove box flushed with pure N_2 (99.9999% purity, no detectable amounts of He, Ne, Kr and Xe by static noble gas mass spectrometry, see Annex 1, Table 3.6).
The pre-crushed sample was put into a silver sample container (length 4.5 cm, diameter 1.3 cm) and loaded into the vacuum gas extraction and purification line. A stainless steel tube that ultimately hosts the silver sample container was then preheated under vacuum to 650°C. Preheating of the sample tube to a higher temperature than the extraction temperature later applied significantly reduced the measured blank signals, as adsorbed noble gases on the inner surfaces were desorbed and pumped away. Next, the sample in the silver container was moved into the preheated section of the stainless steel tube without breaking the vacuum, and was preheated to 100°C to remove adsorbed water from the pre-crushed sample. The extraction of water and noble gases was finally carried out in a 1-h heating step at 300°C - 600°C.

Extraction method II
In order to avoid possible contamination with atmospheric noble gases, which may happen during pre-crushing in air, we built a pestle-operated vacuum crusher that allowed the sample (4-5 pieces of about 0.5 g each) to be crushed in vacuum online, followed by heating, without having to expose the samples to air (an illustration of the vacuum crusher is shown in Annex 1, Figure 3.6).
Prior to crushing, the sample pieces and the vacuum crusher were preheated to 100°C and 300°C, respectively. Water and noble gases were extracted stepwise, with one or more crushing steps, each consisting of 10-50 hammer strokes on the stainless steel pestle of the vacuum crusher (extraction method II-a) being followed by a final heating step at 280°C (extraction method II-b).

3.3.2 Determination of the water mass

Accurate determination of water masses of only a few milligrams is challenging and cannot be accomplished by the standard method used for analysing the noble gas concentrations of water samples, i.e. by weighing the sample before and after sampling[1]. We therefore set up a system (Fig. 3.2) to determine manometrically the mass of the extracted water (see also Kluge et al., 2008 for a similar system). The water liberated from the sample is first frozen at the temperature of liquid nitrogen (approximately -192°C) in a "cold finger" and is then allowed to expand into the calibrated volume V (250 cm^3). V is submerged in a water bath kept at a constant temperature of 40 °C to prevent the water vapour from condensing. A precise pressure gauge measures the water vapour pressure inside V, allowing the water mass to be calculated from the ideal gas law. Water masses of up to ~13 mg can be determined before water starts to condense. An additional volume V* (320 cm^3) can be added to V, giving a combined volume of 570 cm^3 and allowing water masses of up to ~28 mg to be determined. Given the uncertainty of the measured pressure as well as the temperature and the calibrated volume, the overall propagated 1σ-error of the water masses is less than 1.5%.

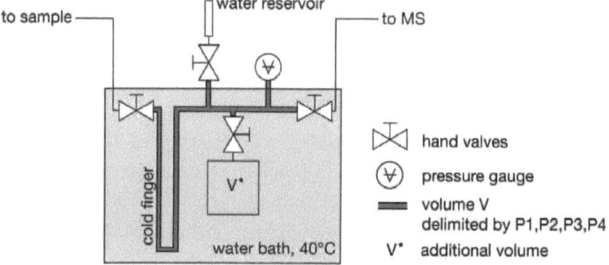

Figure 3.2: Setup of the pressure measurement system. The calibrated volume V contains a cold finger that can be cooled with liquid nitrogen. A pressure gauge (MKS baratron type 626, maximum pressure 100 mbar, precision ± 0.02 mbar), the additional volume V* and a water reservoir are connected to the volume V. The water bath, which is held at a temperature of 40 °C ± 0.5 °C (Lauda Einhängethermostat A100), can be removed to cool the cold finger.

In order to determine the mass of the extracted water in such a manner, the following two requirements have to be fulfilled: 1) the water extracted from the stalagmite sample must be quantitatively trapped in the calibrated volume; and 2) the pressure in the calibrated volume must be generated essentially by the water vapour

Quantitative trapping
To confirm that the extracted water was trapped quantitatively in V, a series of water recovery experiments were conducted. For this purpose we attached a

---

[1] Extensive experiments showed that weighing stalagmite samples before and after noble gas analysis is not feasible. This is because weight changes in the milligram range were not only caused by the loss of water extracted from fluid inclusions, but also by i) loss of grains during the extraction, ii) adsorption of water vapour on the crushed sample during weighing and iii) weight changes of the stainless steel tube and the sample container after heating to high temperatures.

reservoir of degassed water to V (see Fig. 3.2). This procedure allowed a well-defined amount of water vapour to be injected into V. This water was allowed to expand into V only, or into V plus the stainless steel tube, that was either empty or contained a crushed and degassed stalagmite sample. The experiments showed that the water was completely trapped in V when the stainless steel tube and the connection line between the sample and the calibrated volume were heated to 100 °C (Table 3.1).

Table 3.1: Results of the water recovery experiments. The table gives the mass of water injected into the calibrated volume V, the expansion volume and the heating temperature (T) of the sample tube and the connection line to V. The yield refers to the percentage of the initial water that was recovered in the calibrated volume V.

H_2O (mg ± 1.2 %)	Water vapour expanded into	T (°C)	Yield (%)
4.98	volume V	-	100.7
4.89		-	100.0
4.93		-	100.1
4.73		-	100.2
4.71		-	100.0
2.53		-	100.2
1.62		-	100.0
4.92	volume V and sample tube	-	100.6
4.94		-	100.1
4.87	volume V and the sample tube with a crushed sample	0	98.5
5.03		0	99.2
1.24		0	98.2
4.92		0	99.6
5.01		100	100.0
1.10		100	99.7
1.36		100	100.3
1.40		150	99.7
5.35		150	100.3
5.11		150	99.8
4.83	volume V, the sample tube with a crushed sample, water frozen to the sample	100	100.3
4.94		150	99.7
4.97		100	99.8

Pressure by water vapour only

The cold finger in V will trap all gases liberated from the stalagmite sample that reach their freezing point at or above the temperature of liquid nitrogen. This may include atmospheric gases liberated from air inclusions or CO_2 produced by the decomposition of $CaCO_3$ into CaO and CO_2 at temperatures above 600°C (Faust, 1950). By heating completely degassed (gas-free and water-free) stalagmite samples to continuously increasing temperatures, we identified the gas being generated at temperatures > 620°C to be CO_2 using a miniature quadrupole mass spectrometer. As no other gases were detected (except for H_2 in the presence of water, see next paragraph), we conclude that CO_2 does not compromise the water vapour pressure determination at temperatures below 600°C. Also, the pressure of 0.03 mbar resulting from the release of atmospheric gases from a 2 g sample containing 3 vol.% of air can be neglected, as 0.1 wt.% of water released from the same sample already causes a pressure of 11 mbar.

However, during the heating of the stalagmite samples we observed a measurable pressure from a non-condensable gas species, which was identified as hydrogen using the miniature quadrupole mass spectrometer. H_2 first formed at around 300 °C and its partial pressure increased strongly with increasing temperature. Extensive recovery experiments with water from the reservoir showed that water was consumed whenever H_2 was generated[2]. We thus assume that H_2 forms if the water released from the sample reacts with the metal surfaces in the extraction system at temperatures above 300°C. To account for the resulting bias in the water determination, the H_2 partial pressure was used as the error of the pressure reading. For three measurements with extraction temperatures above 400°C this correction increased the error of the determined water mass to 7-10%. However, for the vast majority of measurements the error of the calculated water masses lies below 1.5% (for more information about the water content in samples from 3 stalagmites see Annex 1, Figure 3.7).

3.3.3 Noble gas analysis

The mass spectrometric analysis of noble gases follows the analytical protocol developed for water samples (Beyerle et al., 2000) with modifications required by the ~ 2000 times smaller gas amounts in typical stalagmite samples.

The measured sample signals were calibrated against a standard of about 0.6 cm^3STP of diluted dry air. Standards were measured once a day and treated as common samples. To simulate the analysis of a stalagmite sample we also measured 1) "wet standards" by adding 2-3 mg water from the water reservoir to the standard gas and 2) "heated wet standards", including heating of the stainless steel tube to 600°C for 1 h. All types of standard measurements agreed within experimental error.

Noble gas amounts were calculated by peak height comparison of measured sample signals and the signals of the air standard. The errors of the noble gas amounts account for the individual analytical 1σ-error and the 1σ-standard deviation of the measured standard signals (Table 3.2). After subtracting the blank gas amounts, the noble gas amounts are divided by the manometrically determined water mass to obtain the respective noble gas concentrations. For the vast majority of the analysed samples, the blank correction was not

---

[2] We conducted extensive heating tests with sample containers made of aluminium, nickel and silver. The results showed that the formation of H_2 was strongest for aluminium followed by nickel and silver. Silver hence turned out to the best material for stalagmite sample containers.

significant (< 0.5% of the sample signal). However, for very low sample signals, i.e. very small quantities of extracted water and low amounts of "excess air", blank signals accounted for ~ 1.5% of the sample signal for He to Kr and ~ 30% for Xe (typical blank gas amounts are given in Annex 1, Table 3.7). Overall, noble gas concentrations were determined with analytical errors (median of 1σ errors) of 2.2% for He to Kr and 3.6% for Xe.

Table 3.2: The reproducibility of standards for each noble gas isotope and the isotopic ratios of Ne and Ar, expressed as the 1σ error of measured standard signals over the period of a measurement run.

Isotope	Standard deviation of calibration signals (%)	ratio	Standard deviation of calibration signals (%)
^{40}Ar	1.0	^{20}Ne/^{22}Ne	0.1
^{4}He	0.3	^{36}Ar/^{40}Ar	0.04
^{20}Ne	0.5		
^{86}Kr	0.5		
^{136}Xe	2.1		

3.4 Results and Discussion

We analysed stalagmite samples with ages of 1-50 ka from the Middle East (stalagmites A and B), Turkey (stalagmite C) and central Europe (stalagmites D to H) (Table 3.3[3]). Microscopic analysis of thin sections indicated that stalagmites A to D had high water contents (0.1 - 0.5 wt.%). Stalagmites E to H were from alpine caves and were formed in a more humid climate than stalagmites A to D. The noble gas concentrations determined for the 30 samples analysed are summarised in Table 3.4.

Table 3.3: Information on the stalagmites used in this study[3]. The temperatures listed are measured modern cave temperatures (Fleitmann et al., 2007, 2009), except for those marked with an asterisk, which are estimated temperatures prevailing in the cave at the time of fluid formation (P. Häuselmann, SISKA, La Chaux-de-Fonds, Switzerland, pers. comm. 2008).

Stalagmite	Cave	Country	Altitude (masl)	Age (ka)	Temperature (°C)
A	Dimarshim	Yemen	350	~ 1.5	27
B	Qunf	Oman	650	not dated	27
C	Sofular	Turkey	442	~ 2	13
D	Bunker	Germany	180	~ 7	10
E	Blaettlerloch	Switzerland	390	~ 3	8-9 *
F	Beatus	Switzerland	870	~ 100	8-9 *
G	Beatus	Switzerland	870	~ 50	8-9 *
H	Feés	Switzerland	840	~ 1	8-9 *

---

[3] In the other chapters of this thesis stalagmite A is called D1 and stalagmite C is called SO2.

Table 3.4: Noble gas concentrations in cm^{3}STP per g of water and the mass of extracted water for the analysed stalagmite samples. The errors are given as relative errors in %. The samples with high H$_2$ releases are indicated with an asterisk. The temperature (T) during the heating extraction process and the number of hammer strokes (n) in the crushing step are given for each sample.

Sample	T,n	He (10^{-6})	Ne (10^{-7})	Ar (10^{-4})	Kr (10^{-7})	Xe (10^{-8})	Water (mg)
extraction method I, pre-crushed in air							
A_1	550°C	3.95 ± 2.6%	25.8 ± 2.6%	14.4 ± 2.6%	1.89 ± 2.7%	2.05 ± 3.6%	3.31 ± 2.6%
A_2	400°C	2.95 ± 1.6%	11.2 ± 1.6%	6.36 ± 1.3%	0.99 ± 1.7%	1.21 ± 2.9%	1.74 ± 1.3%
A_3	400°C	3.90 ± 1.8%	8.48 ± 1.9%	4.94 ± 1.6%	0.85 ± 2.1%	1.55 ± 2.9%	1.92 ± 1.6%
A_4	400°C	2.10 ± 2.5%	5.38 ± 2.6%	3.82 ± 2.4%	0.74 ± 2.6%	1.36 ± 3.4%	2.56 ± 2.4%
A_5*	600°C	1.13 ± 7.4%	2.46 ± 7.4%	1.79 ± 7.2%	0.37 ± 7.3%	0.77 ± 7.6%	9.10 ± 7.2%
A_6	400°C	1.71 ± 1.8%	6.79 ± 1.6%	4.41 ± 1.2%	0.82 ± 1.6%	1.81 ± 2.7%	2.33 ± 1.2%
A_7	400°C	2.52 ± 1.9%	8.85 ± 1.9%	5.91 ± 1.2%	1.00 ± 1.9%	1.83 ± 2.7%	1.14 ± 1.2%
A_8	400°C	1.46 ± 1.7%	4.31 ± 1.8%	2.32 ± 1.3%	0.44 ± 1.9%	0.75 ± 3.4%	3.07 ± 1.3%
A_9	600°C	2.26 ± 2.6%	10.2 ± 2.6%	5.82 ± 2.3%	1.04 ± 3.0%	2.02 ± 3.3%	1.97 ± 2.3%
B_1*	400°C	35.6 ± 8.8%	25.2 ± 11%	8.40 ± 8.4%	2.42 ± 13%	7.75 ± 12%	0.40 ± 8.4%
B_2	400°C	8.71 ± 3.0%	20.0 ± 3.4%	8.70 ± 1.9%	1.71 ± 6.0%	3.93 ± 4.5%	0.23 ± 1.9%
C_1*	550°C	10.2 ± 10%	11.3 ± 10%	4.41 ± 11%	1.02 ± 11%	3.29 ± 11%	0.59 ± 10%
extraction method I, pre-crushed in N$_2$							
A_10	320°C	1.31 ± 1.8%	10.6 ± 1.5%	10.5 ± 1.8%	1.10 ± 1.5%	1.38 ± 2.2%	1.99 ± 1.2%
A_11	320°C	2.35 ± 1.5%	9.94 ± 1.4%	20.4 ± 1.8%	1.02 ± 1.5%	1.29 ± 2.9%	1.73 ± 1.3%
A_12	320°C	2.55 ± 1.8%	11.7 ± 1.5%	10.7 ± 1.8%	1.21 ± 1.4%	1.60 ± 2.2%	1.96 ± 1.2%
C_2	320°C	13.4 ± 1.5%	16.0 ± 1.5%	27.4 ± 1.8%	1.46 ± 1.5%	4.06 ± 2.5%	0.23 ± 1.5%
D_1	320°C	29.4 ± 1.7%	11.2 ± 1.7%	7.43 ± 2.0%	1.25 ± 2.0%	3.70 ± 2.7%	0.20 ± 1.5%
H_1	320°C	5.82 ± 1.9%	4.80 ± 1.7%	5.73 ± 1.9%	0.57 ± 1.7%	0.78 ± 2.7%	0.96 ± 1.4%
extraction methods II-a and II-b							
A_13	400°C	0.84 ± 2.1%	20.8 ± 1.9%	10.5 ± 2.1%	1.47 ± 2.0%	1.36 ± 3.6%	0.40 ± 1.8%
A_14	400°C	0.35 ± 2.0%	9.81 ± 1.7%	4.78 ± 1.6%	0.69 ± 1.5%	0.75 ± 3.4%	0.23 ± 1.2%
A_15	10 strokes	7.36 ± 1.4%	172 ± 1.4%	90.6 ± 1.7%	11.9 ± 1.4%	10.1 ± 2.6%	0.67 ± 1.3%
	20 strokes	5.10 ± 7.2%	102 ± 7.2%	50.0 ± 7.2%	7.11 ± 7.1%	6.85 ± 7.8%	0.06 ± 7.1%
	500°C	1.41 ± 2.2%	23.1 ± 2.0%	11.7 ± 2.2%	1.49 ± 2.0%	1.46 ± 3.1%	0.24 ± 1.9%
A_16	10 strokes	19.5 ± 2.0%	457 ± 2.1%	214 ± 2.3%	28.3 ± 2.1%	25.6 ± 3.0%	0.15 ± 2.0%
	20 strokes	9.17 ± 2.1%	195 ± 2.1%	85.7 ± 2.3%	11.6 ± 2.1%	11.4 ± 3.0%	0.15 ± 2.0%
	100 strokes	13.2 ± 3.4%	273 ± 3.4%	119 ± 3.5%	16.7 ± 3.4%	15.6 ± 4.0%	0.06 ± 3.3%
	280°C	0.63 ± 2.5%	12.6 ± 2.5%	4.67 ± 2.4%	0.61 ± 2.6%	0.60 ± 4.9%	0.48 ± 1.3%
C_3	20 strokes	0.38 ± 3.0%	6.74 ± 1.7%	12.8 ± 1.6%	1.42 ± 1.4%	2.01 ± 2.8%	0.28 ± 1.2%
C_4	20 strokes	1.10 ± 2.1%	26.3 ± 2.1%	23.0 ± 2.3%	4.04 ± 2.1%	5.11 ± 3.6%	0.25 ± 2.0%
	500°C	9.44 ± 3.0%	3.66 ± 4.3%	1.85 ± 3.1%	0.44 ± 4.0%	0.44 ± 16%	0.07 ± 2.9%
D_2	10 strokes	1.43 ± 1.3%	43.1 ± 1.3%	24.6 ± 1.6%	3.66 ± 1.3%	3.65 ± 2.5%	1.54 ± 1.2%
D_3	5 strokes	1.64 ± 1.3%	45.5 ± 1.4%	28.1 ± 1.6%	4.17 ± 1.3%	4.33 ± 2.5%	1.06 ± 1.2%
	60 strokes	3.53 ± 1.4%	70.1 ± 1.3%	47.2 ± 1.5%	7.63 ± 1.5%	8.95 ± 3.5%	0.09 ± 1.1%
	280°C	1.49 ± 1.5%	11.4 ± 1.5%	4.76 ± 1.7%	0.54 ± 2.2%	0.41 ± 8.7%	0.16 ± 1.3%
E_1	20 strokes	12.1 ± 1.3%	379 ± 1.4%	190 ± 1.7%	26.4 ± 1.4%	24.2 ± 2.6%	0.31 ± 1.3%
	280°C	3.34 ± 1.3%	67.1 ± 1.3%	38.4 ± 1.6%	5.78 ± 1.4%	6.41 ± 2.5%	0.58 ± 1.2%
F_1	20 strokes	174 ± 10%	3696 ± 10%	1752 ± 10%	214 ± 10%	163 ± 10%	0.02 ± 10%
	280°C	56.9 ± 3.4%	88.9 ± 3.4%	34.9 ± 3.5%	3.70 ± 3.5%	2.30 ± 5.7%	0.06 ± 3.3%
G_1	20 strokes	208 ± 20%	5804 ± 20%	2775 ± 20%	355 ± 20%	297 ± 20%	0.01 ± 20%
	40 strokes	60.8 ± 20%	1411 ± 20%	665 ± 20%	88.4 ± 20%	81.8 ± 20%	0.01 ± 20%
	280°C	4.85 ± 1.5%	13.51 ± 1.6%	6.17 ± 1.8%	0.82 ± 2.1%	0.73 ± 6.7%	0.28 ± 1.4%
H_2	20 strokes	3.87 ± 1.4%	107 ± 1.4%	65.2 ± 1.6%	10.5 ± 1.4%	11.8 ± 2.5%	0.40 ± 1.3%
H_2	280°C	1.71 ± 2.0%	5.56 ± 1.9%	3.70 ± 2.0%	0.64 ± 2.4%	0.88 ± 4.9%	0.17 ± 1.8%

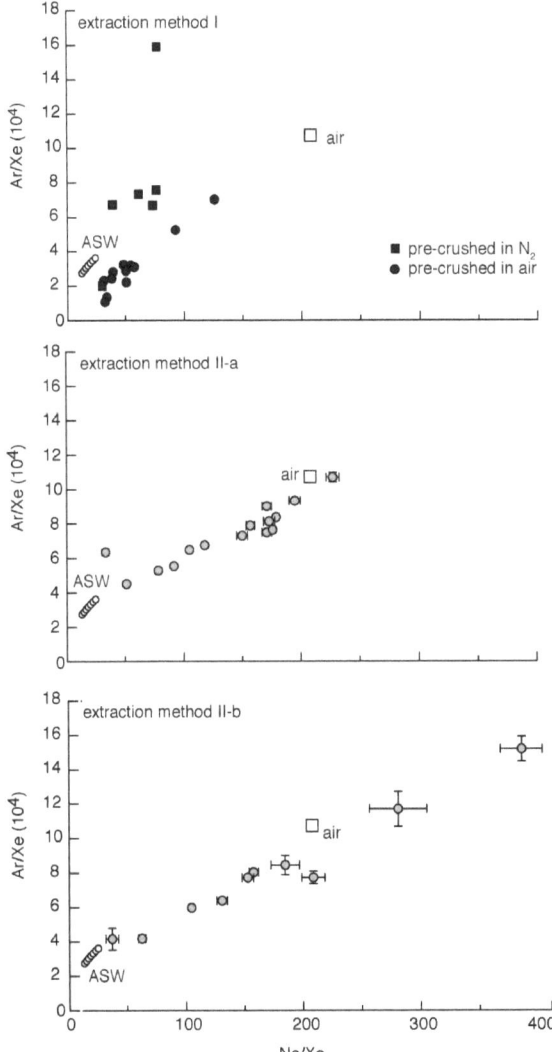

Figure 3.3: Noble gas signature (Ne, Ar and Xe) of the gas extracted from stalagmite samples. The ratios shown were calculated from the measured noble gas abundances. For a better overview, the data resulting from the extraction methods I (dark grey), II-a and II-b (light grey) are shown in three separate plots. The grey shaded area represents mixtures of ASW (0-30°C, open circles on the left) and unfractionated atmospheric air (open square on the right).

3.4.1 Separation of air and water inclusions

In this section we use three-element plots to compare the two extraction methods in terms of their ability to separate noble gases originating from air inclusions from noble gases originating from water inclusions (Fig. 3.3).

In three-element plots, data points representing a mixture of noble gases from atmospheric air and ASW must fall within the area defined by the atmospheric air data point and the array of points given by the temperature-dependent composition of noble gases in ASW. Conceptually, the noble gases extracted from the stalagmite samples can be viewed as a mixture of noble gases originating from ASW and atmospheric air. This agrees with microscopic analyses, which showed that both air and water inclusions are present in the stalagmites. However, it has to be noted that several data points in Fig. 3.3 - especially those associated with extraction method I - fall outside the binary mixing triangle, indicating a possible contribution from one or more additional noble gas components (see section 4.2 and 4.3).

Figure 3.3 shows that the relative abundance of noble gases originating from air inclusions and from water inclusions depends on the extraction method used. The data points from extraction method I generally lie close to the point representing the ASW component and also close to each other, indicating that similar amounts of noble gases from air inclusions are released from these samples. On the other hand, the data points from extraction method II scatter over the whole mixing triangle, i.e. the "excess air" amounts in these samples strongly vary. This indicates that extraction method II does not lead to an adequate separation of noble gases that originate from air inclusions from those that originate from water inclusions. The vacuum crusher hence seems to open air and water inclusions with a similar efficiency. Also, the first few hammer strokes open too many water inclusions (see Table 3.4), leaving only little water to be extracted during the subsequent crushing or heating steps.

The key to a successful separation of noble gases originating from the two types of inclusions thus seems to be pre-crushing and sieving of samples to grains of a defined size given by the size of the air and water inclusions on the one hand and the size of the calcite crystals on the other. Apparently, such size-dependent crushing preferentially opens inter-crystalline air inclusions, while the intra-crystalline water inclusions largely remain intact. As a result, the noble gases released during the heating step originate mainly from water inclusions, leading to a noble gas signature close to the ASW component. For a comparison of grain size distributions of "offline" and "online" crushed samples see Annex 1, Figure 3.8.

3.4.2 Noble gas temperatures

We determined NGTs by error weighted least-squares fitting of the measured concentrations of Ne, Ar, Kr and Xe using the Matlab program noble90 (developed by Aeschbach-Hertig et al., 1999). The measured He concentration was not used to determine NGTs, as large excesses relative to a composition of ASW and atmospheric air were observed in all samples (see later in this section). The least-squares fitting procedure interprets the measured concentrations in terms of the unknown variables T and the volume of dry air per gram of water ("unfractionated excess air"). It minimises the sum of the error-weighted squared deviations between measured and modelled concentrations, denoted by

$$\chi^2 = \sum (C_{i,measured} - C_{i,modelled})^2 / (\Delta C_{i,measured} / C_{i,measured})^2$$

T is then calculated from the ASW component and the "excess air" is a measure of the elementally unfractionated atmospheric air component. The existing least squares fitting models can therefore only be used to interpret binary mixings of air and ASW. Solutions that are statistically consistent with the data are expected to yield χ^2-values corresponding to the degree of freedom in the regression $v = N - P = 2$, where $N = 4$ is the number of noble gases considered (Ne, Ar, Kr, Xe) and $P = 2$ is the number of variables estimated (T, ΔNe).

In our study, least squares fitting of the measured noble gas concentrations in stalagmites (Ne, Ar, Kr and Xe) resulted in χ^2-values larger than the number of degrees of freedom for all samples Therefore, none of our data sets yields a statistically acceptable NGT. This clearly shows that the binary-mixing models implemented in the Matlab code noble90 are conceptually not appropriate to explain the measured noble gas concentrations in stalagmites. We therefore hypothesize that measured noble gas concentrations in stalagmites originate from more than the two atmospheric components that the available methods are able to account for (detailed discussion in section 4.3).

On the other hand, we note that the calculated NGTs for 8 modern stalagmite samples analysed by extraction method I are physically acceptable, as they agree with the present cave temperature to within 1-3 °C. Five samples from stalagmite A (A_1, A_3, A_4, A_6, A_11) with a cave temperature of 27°C yield NGTs between 26 and 30°C, 2 samples of stalagmite C (C_1, C_2) with a present day cave temperature of 13°C result in NGTs of 11-14°C and the NGT of sample D_1 of 10°C agrees with thee present cave temperature of 10°C. The fact that 40% of all samples analysed by method I yield physically meaningful NGTs tends to indicate that stalagmites indeed preserve a temperature signal in their noble gas record and that extraction method I is a very promising experimental protocol. Nevertheless, we again note that even these samples resulted in statistically unacceptably high χ^2-values and correspondingly large uncertainties of 10-30°C (the results of least-squares fitting are shown in Annex 1, Table 3.8).

3.4.3 Noble gas components

In order to understand better why the least squares fitting does not yield satisfying results and to discuss the various geochemical origins of the measured noble gases in stalagmites, we calculated elemental ratios $R = (Ng_i/Ar)_{sample}/(Ng_i/Ar)_{air}$ of the measured noble gas amounts, where Ng_i is the amount of noble gas i. These ratios are used to characterize the deviation of measured noble gas abundances from the noble gas composition of atmospheric air or ASW. We chose Ar as the normalising element as its physical properties lie between those of the more soluble noble gases Kr and Xe and the light noble gases He and Ne, which are good tracers of "excess air" due to their low and only slightly temperature-dependent solubilities.

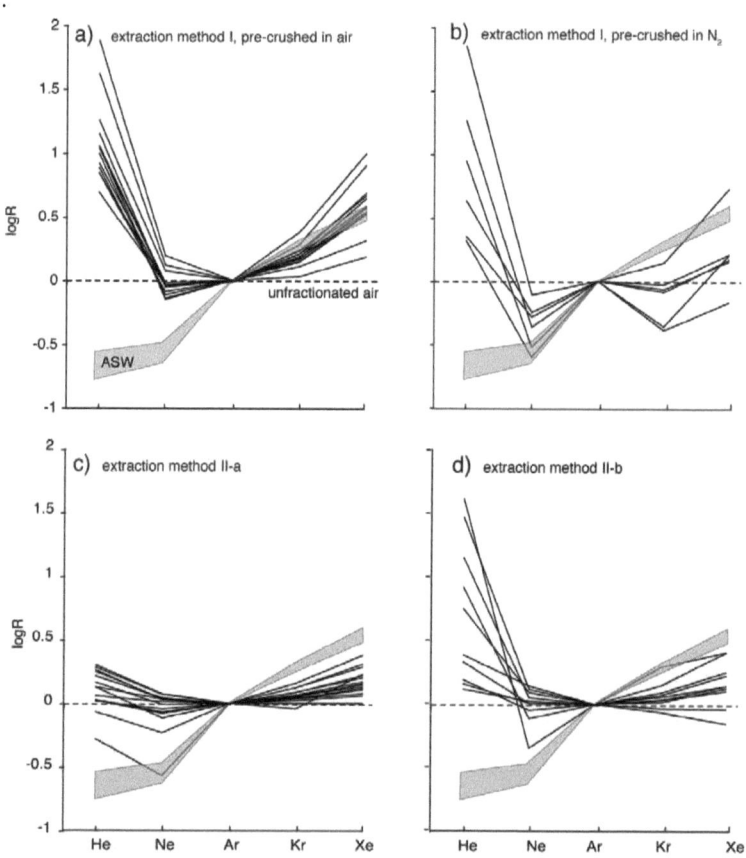

Figure 3.4: Normalised elemental ratios $R = (Ng_i/Ar)_{sample}/(Ng_i/Ar)_{air}$ for each extraction method, expressed as a logarithm. Also shown is the noble gas pattern for air-saturated water from 0-30°C (grey area) and for unfractionated atmospheric air (dashed horizontal line). The region between the grey area and the dashed horizontal line corresponds to a mixture of the two components.

The figure illustrates that Kr and Xe independent of the extraction method employed can be interpreted as a mixture of ASW and different amounts of atmospheric air, as the corresponding lines lie between the grey shaded area representing ASW and the dashed line representing atmospheric air. Exceptions are some samples subjected to extraction method I that were either enriched in Kr or Xe even with respect to the ASW component (Fig. 3.4a) or had lower Kr/Ar ratios than atmospheric air (Fig. 3.4b). As the enrichment with Kr and Xe was only observed in the samples that were in contact with air during crushing, we interpret this enrichment to be an analytical artefact; i.e., to have been produced as a result of atmospheric contamination by adsorption during pre-crushing in air. The noble gas composition of the samples pre-crushed in N_2 support this hypothesis, as the corresponding Kr and Xe excesses are substantially reduced (Fig. 3.4b). However, the N_2 gas we used still contains a substantial amount of Ar (~ 200 ppm) and therefore these samples now show an enrichment in Ar (see also Fig. 3.3). This is very likely the explanation for the fact that the values of R for samples pre-crushed in N_2 lie below the value for atmospheric air as Ar is adopted here as reference element.

In contrast to Kr and Xe, the concentrations of He, and often also of Ne, cannot be explained solely in terms of a simple binary mixing of ASW and air, as R(He) and R(Ne), independent of the extraction method employed, lie outside the region between ASW and atmospheric air. This systematic deviation indicates the presence of an additional noble gas component, which leads to the observed excess in He and Ne relative to the binary mixture of ASW and atmospheric air. Several processes have the potential to enrich He and Ne in stalagmites.

However, most of these can be excluded as a major source of the observed He and Ne excesses. For instance, accumulated radiogenic ^{4}He, which is estimated from the known contents of uranium and thorium and the age of the stalagmite A samples (Fleitmann et al., 2007), accounts only for 10% of the observed He excess. Diffusion of He from the cave atmosphere into the stalagmite cannot cause the observed excess because of the very low diffusion coefficient of He at the ambient temperature and pressure conditions ($D_{He} = 2.4 \times 10^{-25}$ m^2/s: Copeland et al., 2007). Alterations of the equilibrium concentration in water due to a higher partial pressure in the atmosphere in the cave can also be excluded, as the analysis of the noble gas composition of three cave air samples revealed no difference to that of the free atmosphere. We also exclude solubility-related processes, i.e changing solubilities under capillary forces within the water inclusions or in a very thin water film (~ 0.1 mm: Dreybrodt, 1980), as both processes favour only the dissolution of heavy gases relative to light noble gases (Giannesini et al., 2008; Mercury et al., 2004).

Table 3.5: Atomic diameters used in the illustration in Fig. 5 (Momma and Izumi, 2008) and the atomic diameters of noble gases according to Huber et al., (2008).

atom	atomic diameter (Å)	ion	atomic diameter (Å)
He	2.55	Ca	3.94
Ne	2.82	C	1.54
Ar	3.54	O	1.48
Kr	3.65		
Xe	4.05		

Our hypothesis to explain the He and Ne excesses is that these gases are trapped in the calcite crystal lattice (similar processes are discussed by Thomas et al., 2008 and Du et al., 2008). We speculate that He and Ne atoms were trapped in crystallographic voids of the calcite structure (Fig. 3.5) during crystal growth in the cave air. Hence, the lattice-trapped He and Ne are of atmospheric origin. Each crystallographic unit cell of calcite contains 4 such voids in the form of an irregular tetrahedron with a CO_3 group at each corner (Fig. 3.5B). The maximum diameter of a sphere that would fit into the CO_3 tetrahedron void is ~ 3 Å, which is large enough to hold an atom of He or Ne, but not of Ar, Kr or Xe, which have atomic diameters larger than 3 Å (Table 3.5).

Most likely, different processes liberate lattice-trapped He and Ne during crushing or heating. Crushing breaks the calcite crystals predominantly along the cleavage planes of calcite (Miller index {101}) that cuts two of the CO_3 tetrahedron voids in each unit cell. The CO_3 tetrahedron void will then release the He or Ne atom trapped inside. During heating it is more likely that lattice-trapped He or Ne will be released by temperature-enhanced diffusion from the CO_3 tetrahedron void to the surface of the crystal. This also explains the stronger He enrichment relative to Ne observed at higher extraction temperatures, as He is more volatile than Ne (see Fig. 3.4).

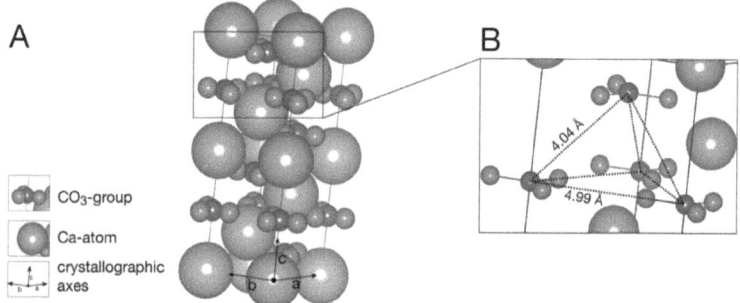

Figure 3.5: A: Schematic diagram of the $CaCO_3$ unit cell with the lengths of the crystallographic axes of a = 4.991 Å, b = 4.991 Å and c = 17.062 Å (Momma and Izumi, 2008). B: Enlargement of the section with the irregular CO_3 tetrahedron void. The radii of the atoms are reduced to 40% of the radii shown in A. The distances between the CO_3 groups are indicated.

Ayliffe et al. (1993) have already speculated about an additional lattice-trapped noble gas component present in stalagmites, besides the noble gases from water and air inclusions. Kluge (2008) observed excesses in Ne relative to a mixture of noble gases from air and ASW in other stalagmite samples, which might also have resulted from the release of lattice-trapped Ne. Moreover, Mohapatra et al. (2009) observed an atmospheric Ne excess in Martian meteorites, which they interpreted as "terrestrial contamination". The meteorite was exposed to terrestrial weathering processes and underwent recrystallization, including the formation of calcite crystals (Ulrich Ott, MPI Chemistry Mainz, Germany, pers. comm. 2009). We therefore speculate that a similar process to the one we propose for stalagmites – i.e., the trapping of atmospheric Ne during the growth of calcite crystals in air – is responsible for the Ne excess observed in weathered meteorites.

3.5 Conclusions and Outlook

The analytical protocol presented in this study enables to determine noble gas *concentrations* in stalagmites precisely, with an overall analytical error of 2-4%. It includes an extraction method (extraction method I) that achieves a sufficient separation of air and water inclusions without loosing too much water during sample preparation. Extraction method I therefore represents an important step forward towards the determination of robust NGTs. For 8 out of totally 18 samples analysed by extraction method I we determined NGTs that correspond reasonably with the present temperature in the cave, even though the NGTs cannot be determined in a statistically acceptable manner. This finding indicates that the available least squares fitting models are not capable to fully explain the observed noble gas concentrations in stalagmites, because they do not account for the additional noble gas components (adsorbed heavy noble gases and lattice trapped He and Ne). As a consequence, the NGTs determined in this study are not robust yet.

According to our results, the additional lattice-trapped noble gas component in stalagmites cannot be extracted separately from the ASW and atmospheric air component, as it is already released during crushing. Also, a quantification of the lattice-trapped component in order to develop new models for least squares fitting seems currently not to be feasible, as this component is up to now only vaguely defined. As a consequence, in the future NGTs have to be determined using only the three noble gases Ar, Kr and Xe for least-squares fitting. Our results indicate that these three gases are not trapped in the crystal lattice, and that their composition is therefore governed only by the two components ASW and atmospheric air if atmospheric contamination during pre-crushing is avoided. As the Ar content in the N_2 gas used for sample pre-crushing in our study is still unacceptably large, we plan to pre-crush future samples in pure He (99.9999%). Also, we intend to improve the extraction method I by defining an "optimal grain size" for pre-crushing for each stalagmite individually, depending on the size of its crystals and of the air and water inclusions. Defining such an "optimal grain size" will further improve the separation of noble gases originating from air inclusions from those originating from water inclusions.

Preliminary results from samples pre-crushed in a He atmosphere are very promising, as the measured concentrations of Kr, Xe and possibly also Ar seem to be determined only by their concentrations in ASW with a small addition of air ($\Delta Ne \sim 50\text{-}100\%$). Hence, least-squares fitting of such binary mixings is expected to yield reliable and robust NGTs in stalagmites, even from a statistical point of view.

The method presented here was developed for stalagmites, but it has the potential to be applied to other samples in which water in the milligram range is present and dissolved atmospheric noble gas concentrations are of interest. For instance, noble gas analysis of the pore waters of consolidated sediments would allow the processes of gas transport through sedimentary basins, and the evolution of these transport processes, to be studied. Biogenic carbonates (e.g., shells and corals), which contain up to 1 wt.% of water (Lecuyer and Oneil, 1994), could also potentially be used to reconstruct the past temperatures of lakes and oceans (Preliminary results of noble gas analysis in a consolidated sediment and in zebra mussel shells are presented in Annex 2, Tables 3.9 and 3.10).

Acknowledegements

The authors would like to thank Philipp Haeuselmann, Marc Luetscher and Tobias Kluge for providing stalagmite samples, Michael Troesch for assisting in the calibration of the pressure measurement system, Christian Baerlocher for his help with the crystallography of calcite, and Henry Schmidt for optimising the pre-crushing procedure. The critical comments of two anonymous reviewers were very helpful to increase the clarity of the manuscript.

3.6 Annex 1

3.6.1 Supplementary Tables

Table 3.6: Ar, Kr and Xe abundance in atmospheric air (Porcelli et al., 2002), in the N_2 gas used to flush the glove box and in the gas phase in the glove box after 10 flushing cycles with N_2. One flushing cycle refers to pumping the chamber until the gloves become inverted and then inflating the chamber with N_2 gas until the gloves protrude out from the chamber. Noble gas abundances are expressed as volume fractions v_i.

	v_{Ar} (-)	v_{Kr} (-)	v_{Xe} (-)
Atmospheric air	$(9.34 \pm 0.01) \times 10^{-3}$	$(1.14 \pm 0.01) \times 10^{-6}$	$(8.70 \pm 0.1) \times 10^{-8}$
N_2 (99.9999% purity)	$(2.1 \pm 0.04) \times 10^{-4}$	$(3.9 \pm 0.2) \times 10^{-11}$	$(1.8 \pm 0.5) \times 10^{-11}$
Gas phase in glove box	$(1.6 \pm 0.01) \times 10^{-3}$	$(1.8 \pm 0.02) \times 10^{-7}$	$(1.4 \pm 0.04) \times 10^{-8}$

Table 3.7: Blank gas amounts of He, Ne, Ar, Kr and Xe measured i) after heating an empty sample tube to 200, 400 and 600°C for one hour, ii) after heating a pre-heated empty sample tube to 300°C and iii) after 100 strokes in the pre-heated vacuum crusher. The last part of the table shows typical noble gas amounts extracted from stalagmite samples for comparison with the presented blank gas amounts.

	He (cm^3STP)	Ne (cm^3STP)	Ar (cm^3STP)	Kr (cm^3STP)	Xe (cm^3STP)
i) Stepwise heating of empty sample tube, no pre-heating					
no heating	3.4×10^{-12}	5.4×10^{-12}	1.6×10^{-9}	2.6×10^{-13}	1.4×10^{-13}
200°C	1.6×10^{-10}	6.3×10^{-10}	1.9×10^{-8}	4.5×10^{-12}	6.6×10^{-12}
400°C	8.5×10^{-12}	1.4×10^{-11}	5.4×10^{-8}	4.8×10^{-12}	7.7×10^{-12}
600°C	6.6×10^{-11}	7.5×10^{-11}	5.2×10^{-9}	6.1×10^{-13}	3.2×10^{-13}
ii) empty sample tube, pre-heated to 400°C					
no heating	7.6×10^{-14}	1.9×10^{-13}	7.0×10^{-11}	1.9×10^{-14}	1.5×10^{-14}
300°C	6.4×10^{-14}	2.6×10^{-14}	1.4×10^{-11}	3.6×10^{-15}	2.1×10^{-15}
iii) vacuum crusher pre-heated to 400°C					
100 strokes	5.4×10^{-14}	8.8×10^{-14}	2.9×10^{-10}	1.4×10^{-14}	4.9×10^{-15}
Noble gas amounts extracted from stalagmite samples					
A_6	3.9×10^{-9}	1.6×10^{-9}	1.0×10^{-6}	1.9×10^{-10}	4.2×10^{-11}
A_16, 10 strokes	2.9×10^{-9}	6.8×10^{-9}	3.2×10^{-6}	4.2×10^{-10}	3.8×10^{-11}
A_16, 280°C	3.0×10^{-10}	6.0×10^{-10}	2.2×10^{-7}	2.9×10^{-11}	2.9×10^{-12}

Table 3.8: The table summarizes the least squares fitting results of all samples analysed by extraction method I and II. Given are the χ^2–values and the two fitted parameters NGT and the amount of "excess air" expressed as % ΔNe. Errors determined by error propagation are scaled with $(\chi^2/\nu)^{1/2}$ to account for the goodness of fit (see Aeschbach-Hertig et al., 1999). Asterisks mark samples that yielded NGTs corresponding within 1-3°C to modern cave temperatures.

	χ^2 (-)	NGT (°C)	ΔNe (10^3 %)
Extraction method I, pre-crushed in air			
A_1*	15.9	26.4 ± 9.9	1.37 ± 0.10
A_2	83.1	36.7 ± 12.0	0.56 ± 0.07
A_3*	250.4	29.6 ± 17.4	0.35 ± 0.10
A_4*	137.1	24.1 ± 10.3	0.18 ± 0.07
A_5	42.4	47.6 ± 17.1	0.04 ± 0.06
A_6*	460.8	28.1 ± 17.1	0.27 ± 0.10
A_7	208.1	19.8 ± 10.7	0.35 ± 0.09
A_8	393.5	66.1 ± 29.6	0.17 ± 0.09
A_9	192.0	17.8 ± 15.0	0.38 ± 0.13
B_1	62.0	0 ± 45.3	0.49 ± 0.61
B_2	265.5	1.4 ± 20.8	0.50 ± 0.28
C_1*	53.5	14.4 ± 30.6	0.26 ± 0.29
Extraction method I, pre-crushed in N_2			
A_10	300.4	22.6 ± 12.3	0.54 ± 0.04
A_11*	1322	26.3 ± 29.8	0.52 ± 0.02
A_12	228.3	18.9 ± 9.9	0.58 ± 0.05
C_2*	1688	11.6 ± 32.1	0.76 ± 0.03
D_1*	486	10.6 ± 15.7	0.42 ± 0.03
H_1	382	38.0 ± 15.3	0.24 ± 0.02
Extraction method II-a and II-b			
A_13	20.0	55.1 ± 17.1	1.31 ± 0.08
A_14	83.0	83.2 ± 26.8	0.71 ± 0.11
A_15	8.7	0 ± 3.3	7.85 ± 0.08
	39.9	1.2 ± 12.1	4.49 ± 0.25
	27.9	63.2 ± 25.8	1.54 ± 0.12
A_16	51.3	0 ± 43.4	20.0 ± 1.2
	119	4.8 ± 38.4	8.61 ± 0.79
	110	0 ± 40.0	11.5 ± 1.0
C_3	739	0 ± 8.6	0.24 ± 0.10
C_4	875	0 ± 26.8	1.18 ± 0.35
	187	96.5 ± 4.1	0.49 ± 0.19
D_2	4.2	2.8 ± 1.7	1.90 ± 0.03
D_3	27.6	0 ± 4.4	1.99 ± 0.09
	497	0 ± 42.7	3.20 ± 0.66
E_1	88.4	0 ± 52.4	17.3 ± 1.3
	88.6	0 ± 14.7	2.99 ± 0.25
F_1	32.4	100 ± 27.3	81.6 ± 5.3
G_1	24.3	0 ± 392	274 ± 10
	63.6	0 ± 180	65.9 ± 4.3
H_2	619	0 ± 59.1	5.36 ± 1.12
	48.9	43.1 ± 8.6	0.28 ± 0.03

3.6.2 Supplementary Figures

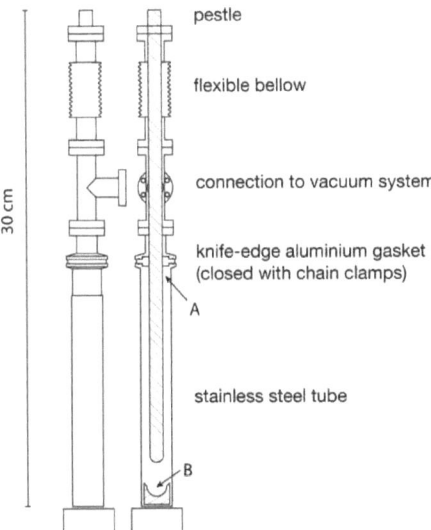

Figure 3.6: Schematic of the vacuum crusher. It consists of a stainless steel pestle, which is connected to stainless steel tube via a flexible bellow. By striking on the top of the pestle with a hammer from outside, the bellow is compressed and allows the pestle to reach the sample and crush it. A: Position of the sample pieces during pre-heating while the vacuum crusher remains in the horizontal position. B: Position of the sample pieces during crushing after turning the vacuum crusher to the vertical position. Fittings are CF flanges with copper seals, except for the connection to the steel tube, which is an aluminium gasket closed with chain clamps.

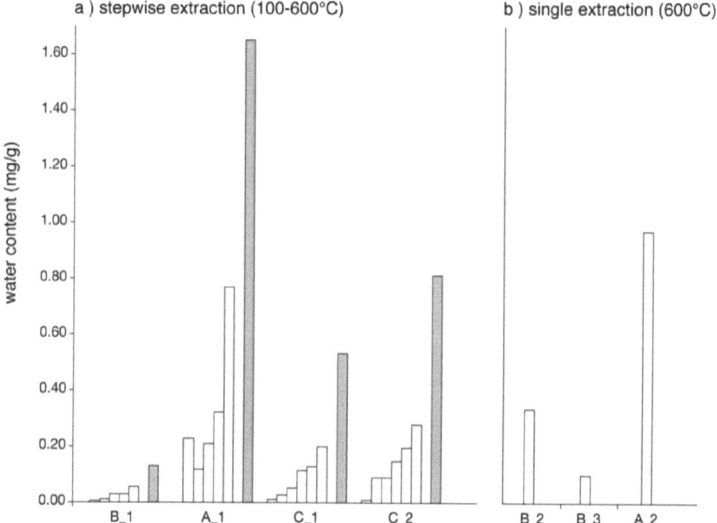

Figure 3.7: Water amounts per gram of sample released from 5 stalagmite samples (stalagmites A, B and C). Samples were pre-crushed in air into grains of 300 μm diameter and then in a) heated from 100 to 600° in 100°C steps of 60 min duration (each white bar represents one heating step) and in b) extracted in a single heating step at 600°C (single white bar). The grey bars represent the totally released amount of water per g of sample. The figure shows that the amount of water extracted from stalagmite samples generally increases with increasing temperature and that around 50% of the totally released water is liberated between 500 and 600°C. However, at temperatures above 300°C part of the released water reacts with the metal surfaces in the extraction system to form H_2 gas. This leads to errors in the determination of the extracted water mass. Therefore, the extraction temperature for stalagmite samples needs to be lower than 300°C.

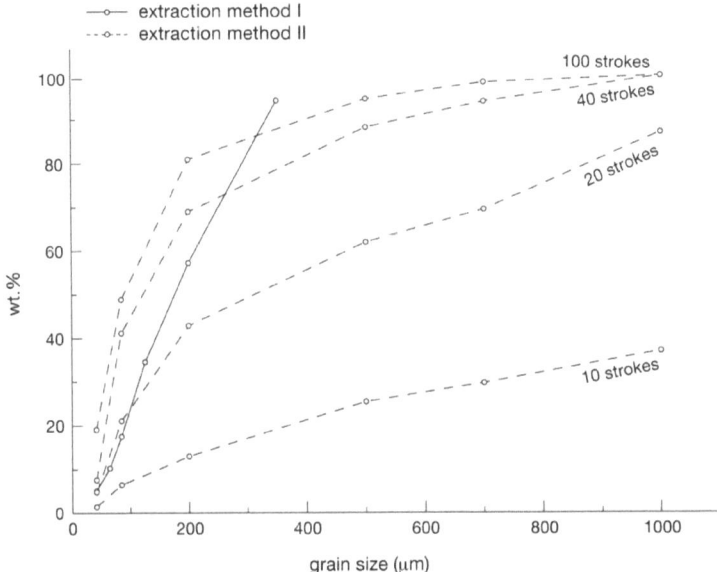

Figure 3.8: Cumulative size distribution of calcite grains from stalagmite D1 after crushing according to extraction method I and II. The Figure shows that with extraction method II the proportion of grains smaller than 300 µm (the grain size of samples crushed with extraction method I) reaches ~100% after 40 to 100 strokes with the vacuum crusher. However, after 40 to 100 strokes only little water remains in the sample to be extracted by heating. As a consequence, extraction method II is not suitable to sufficiently separate air and water inclusions.

3.7 Annex 2

3.7.1 Noble gas analysis in zebra mussel shells

Table 3.9: Noble gas concentrations (in cm^3 STP/g) in water extracted from zebra mussel shells from Lake Constance (Switzerland/Germany). The shells were sampled in water depths of 5m (~ 10°C) and 25m (~ 4°C). Prior to noble gas analysis, the shells were cleaned in an ultrasonic bath and dried for 30 min at 80°C. Noble gases and water were extracted by heating the shells (ca. 1g) for 1h in vacuum to 250-300°C. The error of the extracted water mass is 1.5% and the errors of noble gas concentrations are 1.5% for Ne, Ar, Kr and 3% for Xe.

Noble gas concentrations are expected to be higher in the shells from the deeper water due to the lower temperature. This is reflected in the measured concentrations. However, the measured concentrations are lower than equilibrium concentrations at 4° and 10°C respectively. Also, the concentration difference between the samples of 5m and 25m depth is higher than expected for a temperature difference of 6°C. We suspect that heating of the mussel shells probably released organically bound water, as shells contain up to 5 wt.% of organic material (Mann, 2001). Such "bound water" contains no dissolved noble gases and may hence lead to a virtual lowering of noble gas concentrations. In addition, fractionation caused by the biologically controlled precipitation of the calcium carbonate shell may also affect the noble gas concentrations. We conclude, that noble gas concentrations in mussel shells cannot be interpreted in terms of a water temperature in a simple and straightforward way.

Sample	Water (mg)	He (10^{-8})	Ne (10^{-8})	Ar (10^{-5})	Kr (10^{-8})	Xe (10^{-8})
Samples from 5m						
Shell_1	12.3	6.0	4.6	6.4	2.2	1.4
Shell_2	1.3	1.0	1.8	2.0	1.3	0.6
Sample from 25m						
Shell_3	8.1	5.3	13.9	19.1	7.6	6.3
Equilibrium concentrations in water						
4°C		4.58	20.7	42.4	10.4	1.56
10°C		4.44	19.3	36.7	8.66	1.25

3.7.2 Noble gas analysis in opalinus clay samples

Table 3.10: Amounts of noble gas isotopes per g of sample and the isotopic ratio ^{40}Ar/^{36}Ar measured in samples (~ 0.5g) from the opalinus clay formation at the Mont Terri Rock Laboratory (for details see Pearson et al., 2003). Noble gases were analysed in three extraction steps; 1) degassing at room temperature in vacuum for 14-22 h, 2) online vacuum crushing (50 strokes), 3) heating in vacuum to 250°C for 1h. The errors of the noble gas amounts are 5% for ^{4}He, 3% for ^{20}Ne and ^{40}Ar, 5% for ^{86}Kr and 2-5% for ^{40}Ar/^{36}Ar. For ^{136}Xe the errors are 30% for the samples OPA_1 to OPA_3 and 3% for OPA_4 and OPA_5.

The water content in the samples analysed (~ 1wt.%) is much lower than expected (6-10wt.%, Pearson et al., 2003)). This may be due to diffusive gas and water loss during sampling, sample storage and sample connection to noble gas extraction line. The results are therefore very preliminary. Nevertheless, the results still indicate that different noble gas components are released in the different extraction steps, i.e. the noble gases released in the degassing as well as in the heating step seem to be of atmospheric origin, whereas crushing seemed to be liberating noble gases of radiogenic origin (^{4}He and ^{40}Ar).

Sample	Extraction	^{4}He (cm^{3}STP/g)	^{20}Ne (cm^{3}STP/g)	^{40}Ar (cm^{3}STP/g)	^{86}Kr (cm^{3}STP/g)	^{136}Xe (cm^{3}STP/g)	^{40}Ar/^{36}Ar (-)
OPA_1	degassing	7.4×10^{-11}	2.9×10^{-10}	9.4×10^{-7}	7.9×10^{-10}	4.1×10^{-10}	293.5
	crushing	4.5×10^{-8}	3.3×10^{-10}	6.2×10^{-7}	5.4×10^{-11}	3.0×10^{-11}	927.4
	heating	7.5×10^{-9}	2.2×10^{-10}	1.9×10^{-7}	1.0×10^{-10}	7.9×10^{-11}	293.1
OPA_2	degassing	4.2×10^{-10}	1.3×10^{-9}	8.5×10^{-7}	3.2×10^{-10}	1.4×10^{-10}	301.8
	crushing	8.3×10^{-8}	2.9×10^{-10}	3.9×10^{-7}	3.1×10^{-11}	1.1×10^{-11}	684.2
	heating	1.4×10^{-8}	7.8×10^{-11}	1.0×10^{-7}	6.6×10^{-11}	4.4×10^{-11}	414.9
OPA_3	degassing	1.6×10^{-9}	3.2×10^{-9}	1.8×10^{-6}	5.6×10^{-10}	2.5×10^{-10}	289.0
	crushing	3.9×10^{-8}	5.8×10^{-10}	5.0×10^{-7}	4.0×10^{-11}	1.9×10^{-11}	688.2
	heating	5.4×10^{-8}	1.5×10^{-10}	2.7×10^{-7}	2.0×10^{-10}	1.3×10^{-10}	322.6
OPA_4	degassing	7.3×10^{-9}	1.3×10^{-8}	5.6×10^{-6}	1.3×10^{-9}	6.7×10^{-10}	285.8
	crushing	1.8×10^{-8}	1.8×10^{-9}	6.2×10^{-7}	6.3×10^{-11}	3.5×10^{-11}	353.1
	heating	5.2×10^{-9}	2.0×10^{-10}	2.0×10^{-7}	1.4×10^{-10}	1.6×10^{-10}	282.5
OPA_5	degassing	2.3×10^{-10}	3.3×10^{-10}	6.4×10^{-7}	4.7×10^{-10}	2.9×10^{-10}	293.8
	crushing	5.5×10^{-8}	8.4×10^{-11}	2.9×10^{-7}	3.3×10^{-11}	2.6×10^{-11}	1117.5
	heating	4.4×10^{-8}	1.5×10^{-10}	2.3×10^{-8}	1.4×10^{-10}	1.6×10^{-10}	354.9

4

Determination of Holocene cave temperatures from Kr and Xe concentrations in stalagmite fluid inclusions

This chapter has been submitted to Chemical Geology and is currently in review (Scheidegger et al., in review)

Abstract From the concentrations of dissolved atmospheric noble gases in water, a so-called "noble gas temperature" (NGT) can be determined that corresponds to the temperature of the water when it was last in contact with the atmosphere. Here we demonstrate that the NGT concept is applicable to water inclusions in cave stalagmites, and yields NGTs that are in good agreement with the ambient air temperatures in the caves. We analysed Holocene stalagmite samples from three caves located in three different climatic regions having mean annual air temperatures of 27°C, 12°C and 8°C. In about half of the samples analysed Kr and Xe concentrations originated entirely from the two well-defined noble gas components air-saturated water and atmospheric air, which allowed NGTs to be determined successfully from the concentrations of these two gases. One stalagmite seems to be particularly suitable for NGT determination, as almost all samples analysed from this stalagmite yielded the modern cave temperature. Notably, this stalagmite contains a high proportion of primary water inclusions, which seem to preserve the temperature-dependent signature well in their Kr and Xe concentrations. In future work on stalagmites detailed microscopic inspection of the fluid inclusions prior to noble gas analysis is therefore likely to be crucial in increasing the number of successful NGT determinations.

4.1 Introduction

Stalagmites, which contain precise, high-resolution $\delta^{13}C$ and $\delta^{18}O$ records covering timescales of up to 10^5 years (e.g., Cheng et al., 2009; Fleitmann et al., 2009; Wang et al., 2008), are a major focus of palaeoclimate studies. To explore the full potential of stalagmites as a palaeoclimate archive, direct, independent proxies for cave temperature are needed to constrain and interpret unequivocally the rather complex stable isotope signatures in the calcite, for instance in terms of hydrological variables (Lachniet, 2009). As cave temperatures provide a good measure of the local mean annual air temperature outside the cave (McDermott, 2004), direct cave temperature proxies within stalagmites also allow the reconstruction of palaeotemperatures over long timescales and in continental regions where alternative high-resolution climate archives are often not available.

During stalagmite growth, minute quantities of drip water and cave air are included in the calcite to form fluid inclusions consisting either of air, water or both (Kluge et al., 2008; Scheidegger et al., 2010). Air inclusions account for up to 3% of stalagmite volume (Badertscher, 2007) and water inclusions for 0.01-0.1% of stalagmite weight (Kendall and Broughton, 1978; Scheidegger et al., 2010; Schwarcz and Harmon, 1976). The concentrations of atmospheric noble gases (He, Ne, Ar, Kr, Xe) dissolved in the water inclusions are expected to reflect directly the temperature and salinity of the water as well as the atmospheric pressure at the time of gas exchange with the cave air, as the equilibrium concentrations of noble gases in water are a function of these variables (Kipfer et al., 2002). Hence, noble gas concentrations in stalagmite water inclusions offer the possibility of determining noble gas temperatures (NGTs) if the salinity of the cave water and the atmospheric pressure in the cave are known. The same principle has been widely and successfully applied to groundwater and bulk lake water (Aeschbach-Hertig et al., 2000; Beyerle et al., 1998; Kipfer et al., 2002; Stute et al., 1995; Weyhenmeyer et al., 2000) as well as to pore waters in lake sediments (e.g. Brennwald et al., 2004; Brennwald et al., 2005).

Recent advances in experimental methods now allow noble gas concentrations to be determined even in the small amounts of water (~ 1 mg) typically extracted from a stalagmite sample (Kluge et al., 2008; Scheidegger et al., 2010). However, in these studies, interpretation of the measured noble gas concentrations in terms of NGTs was possible for only a minority of samples. Kluge et al. (2008) were able to convert noble gas concentrations into NGTs solely in stalagmites having an exceptionally low volume ratio of air inclusions to water inclusions. Only when this ratio is sufficiently low can NGTs be determined reliably. This is because noble gases released from air inclusions not only contain no information about the cave temperature, they also mask the temperature-dependent noble gas signature of the water inclusions. To overcome this limitation, Scheidegger et al. (2010) developed a noble gas extraction technique which efficiently separates air inclusions from water inclusions. This method requires samples to be pre-crushed into grains of a defined diameter to remove air inclusions before the noble gases are extracted from the water inclusions by heating. However, during pre-crushing in air and in N_2, the crushed samples adsorbed Ar, Kr and Xe on their freshly created surfaces. This strongly hampered the calculation of NGTs, which were able to be determined for only a few samples and even then with large absolute errors of 10-30°C (Scheidegger et al., 2010).

In this study we analysed noble-gas concentrations in fluid inclusions in stalagmites originating from three different caves located in regions with different annual mean air temperatures (8°C, 12°C and 27°C). The results are based on an improved analytical protocol which reduces noble gas adsorption on the mineral surfaces of the stalagmite samples during the pre-crushing procedure. The measured noble gas concentrations result in precise NGT estimates that are consistent with modern cave temperatures. Hence, this study confirms for the first time that stalagmites are useful archives for reconstructing absolute temperatures using the NGT method.

4.2 Materials and Methods

4.2.1 Samples

We analysed noble gas concentrations in stalagmite samples from Dimarshim cave in Socotra Island, Yemen, Sofular cave on the Black Sea coast of western Turkey and Vallorbe cave, Switzerland (details in Table 4.1). As the caves are located in three different climatic zones, the cave temperatures cover a large part of the temperature range commonly observed in meteoric water systems. The ages of the Dimarshim (D1) and Sofular (SO2, SO3, SO4) cave samples were derived from an age model based on U/Th-series dating (Fleitmann et al., 2007; Fleitmann et al., 2009). No ages are yet available for the Vallorbe cave samples (V1).

Table 4.1: Stalagmites used in this study. The temperatures listed are ambient air temperatures measured in the caves at the time the stalagmite samples were collected.

Sample	Cave	Country	Altitude (m.a.s.l.)	Age of samples (ka)	Temperature (°C)
D1	Dimarshim (12°33'N 53°41'E)	Yemen	350	1.7 to 2.2	27
SO2	Sofular (41°25'N 31°56'E)	Turkey	442	5 to 39	12
SO3	Sofular	Turkey	442	0.1 to 1	12
SO4	Sofular	Turkey	442	2 to 3	12
V1	Vallorbe (46°42'N 6°20'E)	Switzerland	770	not dated	8

4.2.2 Noble gas analysis

Stalagmite samples of 4-6 g were cut along the growth axis of the stalagmite and transferred into a glove box for pre-crushing. The glove box was flushed with 99.9999% pure He which contained no detectable concentrations of Ne, Ar, Kr or Xe (according to our static noble gas spectrometer). Noble gas analysis of gas samples taken in the glove box showed that after extensive flushing Ar, Kr and Xe concentrations were reduced to ~ 10% of their respective concentrations in air (Table 4.2). For further purification, we exposed the gas phase in the glove box for several hours to a sorption pump filled with zeolite (5 Å) and held at the temperature of liquid nitrogen (-196°C). This additional cleaning step reduced the concentrations of Ar, Kr and Xe to ~ 2 % of their respective concentrations in air (Table 4.2).

Table 4.2: Noble gas abundances of Ar, Kr and Xe expressed as volume fractions v_i in the gas phase of the glove box after 20 flushing cycles and after a subsequent exposure of 4h to the sorption pump. Also given are the volume fractions of Ar, Kr and Xe in atmospheric air (Porcelli et al., 2002).

	v_{Ar} (10^{-4})	v_{Kr} (10^{-7})	v_{Xe} (10^{-9})
20 flushing cycles	8.6 ± 0.06	1.2 ± 0.01	9.6 ± 0.4
20 flushing cycles and 4h exposure to the sorption pump	2.0 ± 0.01	0.25 ± 0.002	1.4 ± 0.1
air	93.4 ± 0.1	11.4 ± 0.1	87.0 ± 1

In the glove box, the samples were pre-crushed into grains of a defined diameter to separate air inclusions from water inclusions. Air inclusions tend to be larger than water inclusions and lie preferentially along crystal boundaries. Hence, the preferred diameter of the crushed grains (350 to 700 μm) was chosen based on the size of individual air and water inclusions in each stalagmite, so that air inclusions were preferentially removed during pre-crushing. Because of the preferential removal of air inclusions, the air content in the pre-crushed samples is 100 to 1000 times lower than in bulk samples (Scheidegger et al., 2010). Still in the glove box, the pre-crushed samples were put into a stainless steel tube, which was closed with a vacuum valve. The stainless steel tube was then brought to the noble gas lab and connected to the extraction line without exposing the samples to air.

The noble gases were finally extracted from the pre-crushed samples by heating at 300-400°C for 1h and were analysed using static noble gas mass spectrometry (see Beyerle et al., 2000). Pre-crushing in a He atmosphere led to high He concentrations in the samples. However, this did not influence the analysis of the other four noble gases. Each sample analysis was preceded by the analysis of a blank and followed by the measurement of a standard gas aliquot. Noble gas amounts were calculated by peak height comparison of blank corrected sample signals with the measured standard signal. Dividing the noble gas amounts by the manometrically determined mass of the extracted water (for details see Scheidegger et al., 2010) then yields the noble gas concentrations. These were determined with an overall analytical 1σ error of 2-3% for Ne, Ar and Kr and 3-5% for Xe. The noble gas concentrations determined for the 24 samples analysed are given in Table 4.3.

4.3 Results and Discussion

4.3.1 Noble gas temperatures (NGTs)

In lakes and groundwater the concentrations of Ne, Ar, Kr and Xe usually originate from binary mixtures of two noble gas components, i.e. air-saturated water (ASW) and atmospheric air. This fact allows NGTs to be determined by least-squares fitting of the measured noble gas concentrations (Aeschbach-Hertig et al., 1999; Ballentine and Hall, 1999). However, this approach cannot be applied directly to stalagmites, as not all noble gases released from stalagmite samples have concentrations that are consistent with binary mixtures of two noble gas components. Ne enrichment, for instance, can occur as a result of a lattice-trapped gas component (Scheidegger et al., 2010). Also, in this study we found significant adsorption of Ar on the samples even during pre-crushing in the highly purified gas phase in the glove box.

Table 4.3: Concentrations of Ne, Ar, Kr and Xe in $cm^3 STP/g$ and the mass of the extracted water for the stalagmite samples analysed. Also given are the grain size after crushing and the age of each sample in ka. A typical stalagmite sample of 4-6 g covers a time interval of ~25 a in stalagmite D1, ~ 100 a in stalagmite SO2 and ~ 500 a in stalagmite SO3 and SO4. As stalagmite V1 has not yet been dated, no ages are given in the table for the V1 samples. Samples marked with an asterisk were pre-crushed in the gas phase of the glove box without further purification with the sorption pump.

Sample	Water (mg) ± 1.5%	grain size (μm)	Ne (10^{-7})	Ar (10^{-4})	Kr (10^{-8})	Xe (10^{-9})	Age (ka)
D1_1*	1.39	350	5.92 ± 0.09	4.26 ± 0.15	5.12 ± 0.17	6.21 ± 0.26	1.75
D1_2*	0.98	350	6.85 ± 0.10	4.60 + 0.15	6.02 ± 0.19	7.93 ± 0.31	1.80
D1_3*	1.75	350	6.08 ± 0.09	4.65 ± 0.15	6.13 ± 0.18	7.72 ± 0.29	1.85
D1_4	2.65	350	6.21 ± 0.12	3.38 ± 0.07	5.64 ± 0.12	7.37 ± 0.26	1.90
D1_5	1.84	350	5.74 ± 0.11	3.29 ± 0.07	5.52 ± 0.12	7.32 ± 0.27	1.95
D1_6	1.32	350	6.48 ± 0.12	3.21 ± 0.07	5.19 ± 0.12	6.70 ± 0.27	2.05
D1_7	1.31	350	2.32 ± 0.05	2.19 ± 0.05	3.80 ± 0.09	5.97 ± 0.23	2.10
D1_8	2.59	350	10.5 ± 0.13	5.26 ± 0.06	8.06 ± 0.10	9.24 ± 0.16	2.25
SO2_1*	0.28	500	8.20 ± 0.14	8.00 ± 0.27	9.33 ± 0.29	13.1 ± 0.56	5.1
SO2_2*	0.26	500	6.87 ± 0.13	8.55 ± 0.29	7.99 ± 0.29	11.7 ± 0.66	5.6
SO2_3*	0.27	500	5.54 ± 0.09	5.57 ± 0.09	7.88 ± 0.13	11.3 ± 0.34	6.3
SO2_4*	0.14	500	3.81 ± 0.11	4.84 ± 0.11	7.19 ± 0.18	12.0 ± 0.50	7.1
SO2_5*	0.14	500	5.60 ± 0.15	5.77 ± 0.13	5.99 ± 0.20	15.4 ± 0.58	10.5
SO2_6	0.48	600	1.30 ± 0.04	2.46 ± 0.04	2.52 ± 0.05	5.83 ± 0.19	37.6
SO2_7	0.64	1000	1.65 ± 0.05	2.39 ± 0.06	2.87 ± 0.07	6.57 ± 0.21	38.6
SO3_1	0.35	600	4.24 ± 0.09	3.75 ± 0.05	6.20 ± 0.10	9.11 ± 0.27	0.5
SO4_1	1.12	700	2.85 ± 0.06	3.33 ± 0.07	6.67 ± 0.13	11.58 ± 0.25	2.5
V1_1	0.37	350	8.62 ± 0.18	5.86 ± 0.13	10.47 ± 0.25	15.56 ± 0.71	
V1_2	0.54	350	5.07 ± 0.11	5.44 ± 0.12	9.54 ± 0.22	15.77 ± 0.64	
V1_3	0.92	700	11.49 ± 0.39	6.76 ± 0.23	10.74 ± 0.36	16.14 ± 0.58	
V1_4	1.06	700	6.34 ± 0.08	5.99 ± 0.07	11.37 ± 0.14	16.36 ± 0.28	
V1_5	0.78	600	6.71 ± 0.13	3.83 ± 0.07	6.11 ± 0.12	12.96 ± 0.30	
V1_6	0.93	500	3.60 ± 0.06	3.61 ± 0.05	7.08 ± 0.11	9.62 ± 0.77	
V1_7	0.46	500	2.71 ± 0.05	3.11 ± 0.04	5.85 ± 0.10	9.15 ± 0.80	

In contrast to Ne and Ar, the concentrations of the most soluble noble gases, Kr and Xe, can be interpreted well in about half of the samples analysed as binary mixtures of ASW and atmospheric air (see Figure 4.1). The concentrations of Kr and Xe in these samples lie either directly on, or slightly above, the line representing the temperature-dependent ASW component, indicating the presence of only small amounts of "excess air". The results from these samples also demonstrate that the gas phase in the glove box is sufficiently pure to reduce the amount of Kr and Xe adsorbed on crushed calcite grains to a negligible level. As a result, the Kr and Xe concentrations in these samples are identified as simple binary mixtures of ASW and atmospheric air only and can hence be converted in a straightforward manner into temperature information.

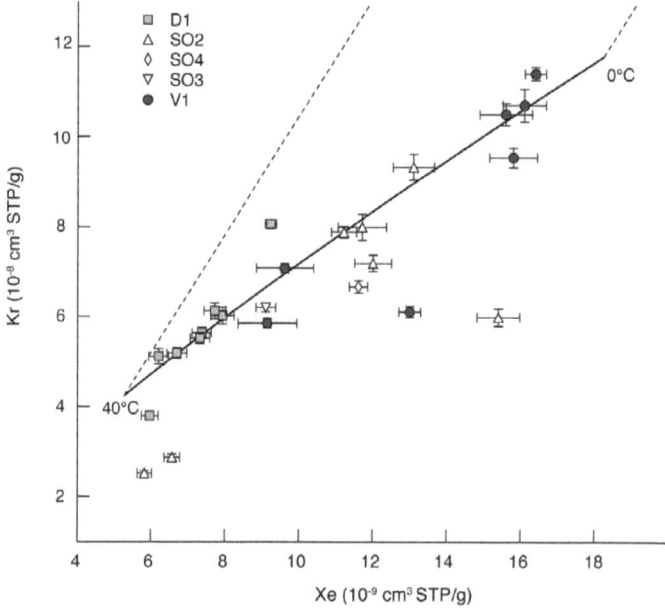

Figure 4.1: Kr and Xe concentrations measured in the stalagmite samples. The solid line represents ASW from 0 to 40°C and the dashed lines indicate the addition of unfractionated "excess air" to the ASW component. Mixtures of ASW and atmospheric air hence lie on lines lying parallel to the dashed lines.

We note that such an interpretation does not apply to the other half of the samples, whose data points lie below the ASW line in Figure 4.1. In these samples, Kr and Xe concentrations are present in proportions, that cannot be explained by a simple binary mixture of ASW and atmospheric air. We speculate that such "strange" Kr and Xe concentrations are related to the calcite structure of a stalagmite and the origin of its fluid inclusions (see section 4.3.2). For the samples with Kr and Xe concentrations on or above the ASW line in Figure 4.1, NGTs were calculated by solving the following simultaneous equations for the temperature (T) and the amount of "excess air" per gram of water (A).

$$C_{Kr} = C^*_{Kr}(T, S, p) + A\, z_{Kr}$$
$$C_{Xe} = C^*_{Xe}(T, S, p) + A\, z_{Xe}$$

C_i is the measured Kr or Xe concentration, C^*_i the respective ASW concentration (depending on the water temperature T, the salinity S and the atmospheric pressure p), and z_i the volume fraction of Kr or Xe in atmospheric air (Porcelli et al., 2002). The salinity of the drip water was assumed to be negligible (i.e. S = 0 g/kg). The partial pressures of Kr and Xe in cave air were determined using the barometric formula to calculate the atmospheric pressure at the elevation of the cave site under the assumption that the noble gas composition of the cave air was the same as that of the atmosphere.

A possible caveat to this assumption is that CO_2 concentrations of several percent have been measured in air inclusions in stalagmites (Badertscher, 2007). This indicates that a gas layer with elevated CO_2 concentrations may exist around the stalagmite, where CO_2 produced during calcite precipitation tends to accumulate. If so, the partial pressures of Kr and Xe in this gas layer would be lowered to the same extent (e.g. 10% of CO_2 reduces the partial pressures of Kr and Xe by 10%), leading to reduced equilibrium concentrations in the fluid inclusion water. This would bias NGTs derived from such samples towards higher temperatures.

Table 4.4: Results of the NGT calculation for all samples with noble gas concentrations that can be conceptually explained with a binary mixture of ASW and atmospheric air. Listed are the calculated NGT and the amount of "excess air" (A).

Sample	NGT (°C)	A (10^{-3} cm^3/g)
D1_1	36.4 ± 3.9	4.3 ± 4.0
D1_2	25.3 ± 3.0	1.1 ± 0.4
D1_3	27.9 ± 3.2	5.2 ± 4.3
D1_4	27.6 ± 2.7	0.6 ± 0.3
D1_5	27.3 ± 2.6	0
D1_6	31.0 ± 3.3	0.2 ± 0.3
D1_8	27.5 ± 1.7	21.6 ± 2.3
SO2_1	9.7 ± 2.3	5.5 ± 0.6
SO2_2	11.3 ± 3.2	0
SO2_3	12.6 ± 1.6	0
V1_1	3.4 ± 2.0	3.2 ± 6.4
V1_3	2.5 ± 1.7	2.3 ± 6.4
V1_4	3.0 ± 0.8	10.3 ± 3.1
V1_6	17.4 ± 5.3	2.6 ± 7.5

Table 4.4 shows the results of the NGT calculation, under the assumption that gas exchange occurred with air of atmospheric composition. The uncertainties in T and A were calculated by Monte-Carlo propagation of the analytical errors of the Kr and Xe concentrations. As all samples were deposited during the Holocene, we assume that no major temperature changes occurred in the cave regions during the time interval covered by our samples. Calculated NGTs are therefore expected to correspond closely to the modern cave temperatures. In 7 samples from stalagmite D1 and in 3 samples from stalagmite SO2, the NGTs correspond within their 1σ errors to the respective modern cave temperatures of 27°C and 12°C (Figure 4.2). The mean NGT and the standard deviation of

the SO2 and D1 samples are 11.2 ± 1.9°C and 29.0 ± 3.7°C, respectively, which are in excellent agreement with the modern cave temperatures. This shows that for most samples, the possibility of elevated CO_2 concentrations around the stalagmite during gas exchange with the fluid inclusion water can be neglected. One sample in stalagmite D1, however, yields an NGT considerably higher than the expected cave temperature. This might be explained by reduced partial pressures of Kr and Xe during air-water partitioning due to an elevated CO_2 concentration (20% CO_2 in the air layer would reduce Kr and Xe partial pressures by 20% and result in an NGT of 27°C for this sample).

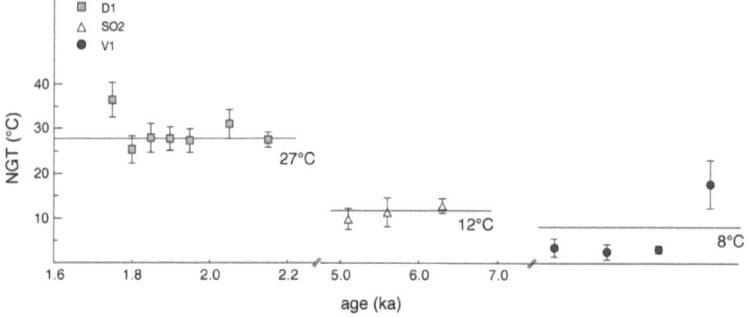

Figure 4.2: NGTs determined from the measured Kr and Xe concentrations plotted against sample age and in comparison with the temperature measured in the caves at the time of sample collection. As stalagmite V1 has not yet been dated, samples are shown in stratigraphic order.

The calculated NGTs of three samples from stalagmite V1 are almost identical to each other, whereas a fourth sample yields a higher temperature with a large uncertainty due to lower analytical precision in the determination of the Kr and Xe concentrations. This latter sample is therefore regarded as an outlier and will not be discussed further. The mean NGT determined from the three V1 samples (3.0 ± 0.5°C) is about 5°C lower than the cave temperature measured at the time of sample collection in May 2009 (8°C). The local mean annual air temperature in the region outside the cave is 7-8°C (data from the Swiss Federal office of Meteorology and Climatology). However, because a large subsurface river flows through Vallorbe cave, the ambient air temperature in the cave does not remain constant throughout the year as it does in many other caves. The calculated NGT in the three Vallorbe cave samples could therefore indicate that growth of the stalagmites in Vallorbe cave occurred mainly during the cold part of the year resulting in an NGT below the mean annual air temperature. Alternatively, the NGT of 3.0°C may also be interpreted as palaeotemperature of a pre-Holocene period, as stalagmite V1 is not dated and was not actively growing when sampled in 2009.

4.3.2 Suitability of stalagmites for NGT determination

We have shown that Kr and Xe concentrations in about half of the samples analysed in this study correspond to the ambient temperature in the cave, as the calculated NGTs are reproducible and in good agreement with modern cave temperatures. However, not all stalagmites studied yielded the same high proportion of samples that allowed successful NGT determination. Almost all samples analysed from stalagmite D1 (7 out of 8) had Kr and Xe concentrations consistent with binary mixtures of ASW and atmospheric air. This interpretation applies to only 3 out of the 9 samples analysed from the Sofular cave stalagmites (SO2, SO3, SO4) and to only 3 out of the 7 samples from the Vallorbe cave stalagmite (V1). Hence, the temperature information seems to be better preserved in the fluid inclusions of stalagmite D1 than in the fluid inclusions of the other stalagmites used in this study.

In order to better understand why some stalagmites are more suitable for NGT determination than others, we analysed the mineralogy of the stalagmites D1, SO2 and V1 using optical and electron microscopy. Optical microscopic inspection of thick sections (~100 μm) showed that large, elongated air inclusions are very abundant in stalagmites SO2 and V1 but are not present in stalagmite D1 (Figure 4.3 D-F). We speculate that these large, elongated inclusions in stalagmites SO2 and V1 were formed by crystal coalescence (Kendall and Broughton, 1978), as they are aligned along the boundaries of neighbouring columnar calcite crystals. Jo et al. (2010) observed similar inclusions in a stalagmite from a cave in Korea, and speculate that these inclusions were formed by coalescence of neighbouring crystals. Crystal coalescence could lead to the formation of pseudo-secondary inclusions, as it occurs after the initial calcite crystals are precipitated (Kendall and Broughton, 1978). Crystal coalescence may also involve secondary gas and pore-water exchange processes, which are likely to affect the noble gas signature in the initially trapped fluid inclusion water. The noble gas signature of such pseudo-secondary inclusions is then no longer suitable for NGT determination. The absence of large, elongated inclusions in stalagmite D1 implies that no crystal coalescence occurred there. This indicates that fluid inclusions in stalagmite D1 are mainly of primary origin and that dissolved Kr and Xe concentrations still reflect the environmental conditions prevailing in the cave at the time of fluid entrapment.

Kr and Xe concentrations may also be altered during the analytical procedure. SEM imaging of fractures parallel to the growth axis of the stalagmites showed that stalagmites SO2 and V1 are comprised of a very dense, homogenous calcite with only few inclusions (Figure 4.3 A-B). We hypothesize that the calcite crystals in such samples often break up into sub-crystals of individual crystals during pre-crushing. This could lead to tiny defects and fissures along the fractures that may connect up to fluid inclusions making them "leaky". As a consequence, Kr and Xe gases may be lost. This could occur, for example, by diffusion along the tiny fissures during pre-crushing in the glove box, in which the gas phase is virtually free of Kr and Xe. In the fast-growing (~300 μm/a) stalagmite D1 the calcite crystals are much smaller and are arranged more irregularly than in stalagmites SO2 and V1 (Figure 4.3, C). The relatively loose arrangement of calcite crystals in stalagmite D1 indicates that the calcite in such samples predominantly breaks along crystal boundaries during pre-crushing, so that individual crystals, and the water inclusions trapped inside them, remain intact. The noble gas concentrations in the fluid inclusions of stalagmite D1 are therefore not affected by alteration processes during the analytical procedure, thus allowing the successful determination of NGTs from the D1 samples.

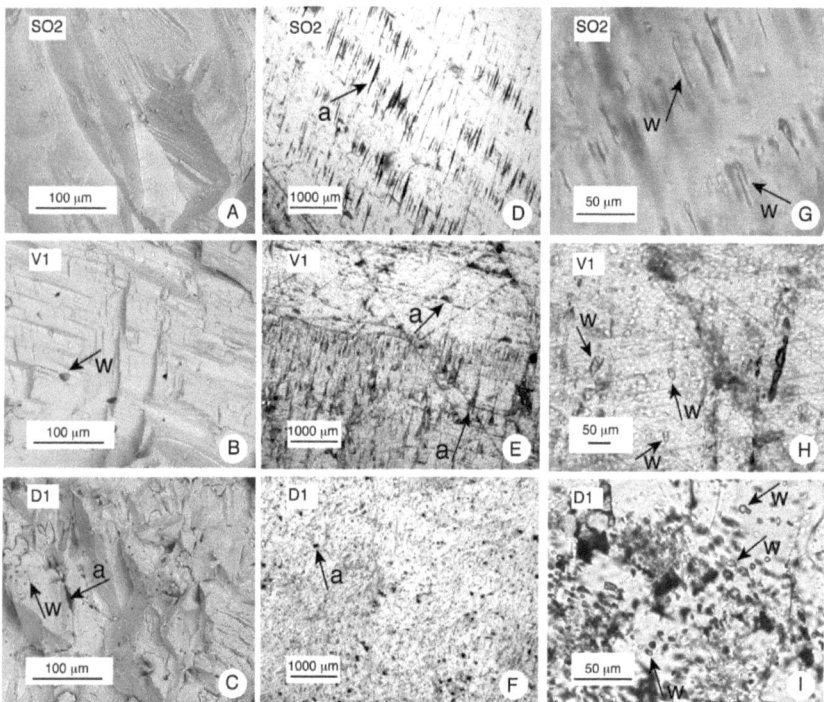

Figure 4.3: Petrographic information on the stalagmites used in this study. The arrows indicate air inclusions (a) and water inclusions (w). **A-C)** SEM images showing the calcite structure of fresh fractions in stalagmites SO2, V1 and D1. **D-F)** Photographs from thick sections taken using an optical microscope at low magnification in cross-polarized light. **G-I)** Photographs from thick sections taken using an optical microscope showing typical water inclusions found in stalagmites SO2, D1 and V1.

4.4 Conclusions and Outlook

This study provides for the first time direct experimental evidence that stalagmite fluid inclusions provide a suitable archive for the accurate and precise determination of past NGTs. Our results demonstrate that NGTs are recorded reliably in water inclusions, as they correspond well with modern cave temperatures in stalagmites D1 and SO2 and are reasonably well related to today's cave temperature in stalagmite V1.

Such absolute determinations of cave temperatures are important for reducing the uncertainties associated with the interpretation of stable isotope records in the stalagmite calcite. Given the current accuracy of NGTs (1-4°C) and the amount of sample currently needed (4-6 g), the reconstruction of major climate changes such as the temperature shifts during the Pleistocene-Holocene transition or during the termination of other glacial cycles (Cheng et al., 2009; Jouzel et al., 2007; Petit et al., 1999), becomes feasible.

Our study also showed that not all stalagmites are equally suitable for NGT determination. The NGT information seems to be better preserved i) in fast growing stalagmites with relatively small, irregularly arranged calcite crystals, and ii) in stalagmites with a high proportion of primary fluid inclusions (e.g. in stalagmite D1). In other stalagmites analysed in this study, the Kr and Xe concentrations seemed to have been altered, possibly by secondary processes occurring during stalagmite growth (e.g. crystal coalescence) or during the noble gas analysis procedure. This prevented the calculation of NGTs in about half the samples analysed from such stalagmites. A detailed microscopic investigation of the calcite structure and of the origin of the fluid inclusions is therefore an important prerequisite for sample selection prior to noble gas analysis. This should result in an improvement in the number of successful NGT determinations able to be conducted on stalagmites in the future.

Acknowledgements

The authors would like to thank Urs Menet and Heinrich Baur for their assistance in the laboratory, and Henry Schmidt for improving the crushing system and for the noble gas analyses he performed during his master's thesis. We also thank Sebastian Breitenbach for the idea of using He to flush the glove box. Thanks also go to David M. Livingstone for proof-reading the manuscript, and to four anonymous reviewers who helped to improve an earlier version of the manuscript.

5

Application of the noble gas thermometer to fluid inclusions in two Holocene stalagmites from Socotra Island (Yemen)

This chapter is in preparation for publication in Geology.

Abstract The concentrations of dissolved noble gases in water are widely used to determine the noble gas temperature (NGT) of the water, i.e. its temperature at the time of the last gas exchange with the atmosphere. Recent analytical advances now allow this principle to be applied to fluid inclusions in stalagmites in order to determine cave temperatures prevailing during fluid formation. In this study, NGTs are determined from Kr and Xe concentrations measured in water inclusions in two stalagmites from Socotra Island (Yemen) covering the last 4.5 and 10 ka. In addition to the NGTs, we also present records of the specific water content and the amount of "excess air" in the analysed samples, which seem to have the potential to be used as new potential climate proxies in stalagmites. In both stalagmites, the calculated NGTs scatter considerably more than would be expected from actual temperature variations during the Holocene in a low latitude region such as Socotra Island. We hypothesize that the variation may be caused by reduced partial pressures of Kr and Xe during air-water partitioning due to an occasional gas layer around the stalagmite with an air like composition but elevated CO_2 concentrations. As a consequence, the conditions during atmospheric noble gas partitioning in fluid inclusions in stalagmites must be understood in more detail, particularly the CO_2 concentration should be quantitatively measured in order to accurately and reliably determine NGTs in stalagmites in the future.

5.1 Introduction

Dissolved atmospheric noble gas concentrations in water can be used to determine the noble gas temperature (NGT), i.e. the temperature of the water when it was last in solubility equilibrium with atmospheric air. NGT determination is based on the temperature dependent solubilities of noble gases in water (Aeschbach-Hertig et al., 1999; Kipfer et al., 2002) and has been ssuccessfully used to infer palaeoclimate conditions from noble gas concentrations in groundwater (e.g. Beyerle et al., 1998; Stute et al., 1995a,b; Weyhenmeyer et al., 2000) in sediment pore waters (Brennwald et al., 2004; Brennwald et al., 2005) and recently also in water-filled fluid inclusions in stalagmites (Kluge et al., 2008; Scheidegger et al., 2010; Scheidegger et al., in review).

Stalagmites represent an increasingly studied palaeoclimate archive, because they can be precisely dated and provide information about continental palaeoclimate conditions over very long time intervals of up to several 10^5 years (e.g. Cheng et al., 2009; Wang et al., 2008; Zhao et al., 2010). Also, caves and stalagmites are found in karst regions all over the world including those regions where usually no other continental climate archives covering glacial-interglacial timescales are present (e.g. Pons-Branchu et al., 2010; Vaks et al., 2010; Spötl and Mangini, 2002). The most widely used climate proxy in stalagmites is the oxygen isotope composition of the calcium carbonate. $\delta^{18}O_{calcite}$ values can be determined with a very high precision and with a high temporal resolution. However, $\delta^{18}O_{calcite}$ reflects both the isotope composition of the drip water ($\delta^{18}O_{dripwater}$) and the temperature of calcite precipitation. Hence, for a sound interpretation of the $\delta^{18}O_{calcite}$ records, independent cave temperature estimates are needed to obtain a quantitative $\delta^{18}O$ record of the drip water, which is closely related to the isotope composition of local precipitation (McDermott et al., 2005).

Dissolved noble gas (Ne, Ar, Kr and Xe) concentrations in water-filled inclusions in stalagmites represent such a direct and independent cave temperature proxy, which allows NGTs to be determined based on the well-described physical process of gas exchange (Aeschbach-Hertig et al., 1999; Kipfer et al., 2002). NGTs in stalagmite fluid inclusions hence offer the possibility of interpreting the $\delta^{18}O_{calcite}$ records more quantitatively and unequivocally in terms of climatologically relevant parameters (e.g. the amount and the source of the precipitation). Kluge et al. (2008) determined the temperature difference between the early and late Holocene in a stalagmite from Germany using atmospheric noble gas concentrations in fluid inclusions. Scheidegger et al. (in review) successfully determined NGTs from Kr and Xe concentrations in various modern stalagmite samples from caves with different mean annual air temperatures.

In this study we analyse noble gas concentrations in two Holocene stalagmites from Socotra Island with the goal to determine past climate conditions and NGTs for the time interval covered by the two stalagmites. The stalagmites were selected using light and electron microscopy based on criteria for sample selection defined by Scheidegger et al. (in review). In both stalagmites, the calcite crystals are irregularly arranged and water inclusions are found within calcite crystals and are therefore most likely of primary origin (Figure 5.1). The two stalagmites are hence expected to be suitable for NGT determination.

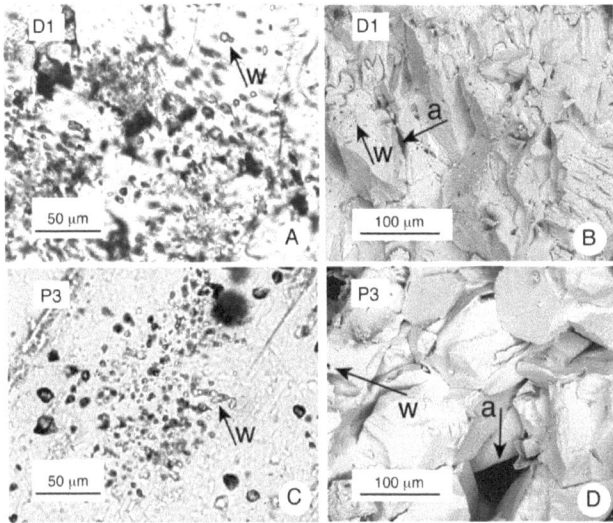

Figure 5.1: Petrographic information on the stalagmites (D1 and P3) used in this study. The figure includes for each stalagmite a photograph from a thick section taken using an optical microscope (A and C) and a SEM image (B, D) to show the calcite structure. The arrows indicate air inclusions (a) and water inclusions (w).

Our results show that the NGTs calculated from Kr and Xe concentrations are so far not adequate and precise enough to be used to constrain the $\delta^{18}O_{calcite}$ records of the stalagmites, because the NGTs show a considerably larger variation than could be attributed to actual temperature changes on Socotra Island during the Holocene. A potential explanation for this variation is presented and discussed in this paper. We further introduce the water content and the amount of "excess air" in the samples are introduced as potential new climate proxies in stalagmites, which may provide valuable complementary information for the interpretation of $\delta^{18}O$ records.

5.2 Climate on Socotra Island

Socotra Island is part of the Republic of Yemen and is situated in the north-western Indian Ocean, about 240 km east of the Horn of Africa and 380 km south of the Arabian Peninsula. The climate on the Island is strongly influenced by the seasonally reversing Indian Ocean Monsoon wind system (Figure 5.2). The associated passing of the Inter-tropical convergence zone (ITCZ) over Socotra in May-June and September-November leads to a bimodal distribution of rainfall on Socotra Island with a maximum in November (Shakun et al., 2007; Fleitmann et al., 2007; Scholte and De Geest, 2010).

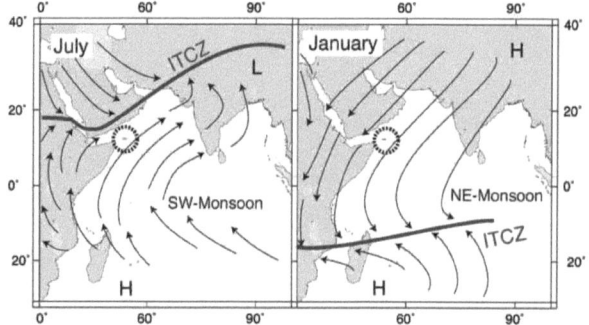

Figure 5.2: Schematic figure of the surface wind patterns in the broader Socotra Island (dotted circle) region during the SW-Monsoon in July and the NE-Monsoon in January. The bold grey line depicts the modern location of the ITCZ (adapted from Fleitmann et al., 2007).

5.3 Material and Methods

We analysed samples from two Holocene stalagmites (D1 and P3), which were collected in two caves from the central part of Socotra Island (Figure 5.3). Stalagmite D1 is from Dimarshim cave (12°33'N 53°41'E), that lies at 350 masl. Stalagmite P3 was collected in Pit cave (12°25'N 53°58'E) located at 480 masl on the limestone plateau south of the central mountain range of Socotra Island. The mean annual air temperature at both cave sites, estimated from modern temperature measurements (Scholte and de Geest, 2010), is ~ 28°C. As no temperature monitoring data is available for Dimarshim and Pit cave, we assume that the temperatures in Pit and Dimarshim cave reflect mean annual air temperatures in the cave region and remain stable throughout the year (McDermott et al., 2005).

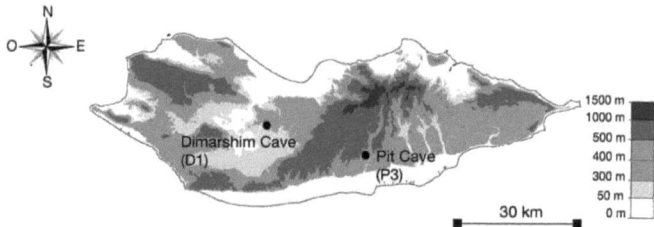

Figure 5.3: A Map of Socotra Island showing the locations of Dimarshim and Pit cave. The elevations are based on data collected in the Shuttle Radar Topography Mission (SRTM).

Totally, we analysed 30 samples from stalagmite D1 covering the last 4.5 ka and 13 samples from stalagmite P3, which cover the last 10 ky. The age of the samples was determined by linear interpolation between absolute U/Th-ages, which are available for both stalagmites (8 for stalagmite D1 from Fleitmann et al. (2007) and 2 for stalagmite P3). The temporal resolution of the samples is ~ 30 years for the D1 stalagmite and ~ 100 years for stalagmite P3.

Table 5.1: Concentrations of Ne, Ar, Kr and Xe in cm^{3}STP/g and the water content in mg of water per g of sample for the stalagmite samples analysed. The Ne and Ar concentrations are given for completeness although they are not used for NGT calculation. Kr and Xe concentrations in samples marked with an asterisks (*) could not be converted into an NGT as the measured concentrations are lower than expected for solubility equilibrium. The data of 7 samples from stalagmite D1 are from (Scheidegger et al., in review) and are marked with †.

Sample	Age (ka)	Ne (10^{-7})	Ar (10^{-4})	Kr (10^{-8})	Xe (10^{-9})	Water content (10^{-1})
D1_1	0.04	30.9 ± 0.4	18.6 ± 0.3	26.5 ± 0.4	28.8 ± 0.6	1.96 ± 0.02
D1_2	0.07	26.0 ± 0.4	11.0 ± 0.2	14.8 ± 0.2	12.5 ± 0.3	2.58 ± 0.03
D1_3	0.19	14.4 ± 0.2	8.48 ± 0.11	12.5 ± 0.2	15.2 ± 0.3	1.77 ± 0.02
D1_4	0.30	30.5 ± 0.4	16.8 ± 0.2	22.0 ± 0.3	20.1 ± 0.4	2.22 ± 0.03
D1_5	0.37	16.7 ± 0.2	8.82 ± 0.11	10.5 ± 0.2	10.5 ± 0.2	2.53 ± 0.03
D1_6	0.74	5.63 ± 0.08	3.49 ± 0.05	5.49 ± 0.08	6.62 ± 0.15	3.31 ± 0.04
D1_7	1.60	3.77 ± 0.05	2.68 ± 0.04	4.86 ± 0.07	4.69 ± 0.10	3.02 ± 0.04
D1_8†	1.75	5.92 ± 0.09	4.26 ± 0.15	5.12 ± 0.17	6.21 ± 0.26	4.96 ± 0.08
D1_9	1.80	6.85 ± 0.10	4.60 ± 0.15	6.02 ± 0.19	7.93 ± 0.31	3.50 ± 0.04
D1_10†	1.85	6.08 ± 0.09	4.65 ± 0.15	6.13 ± 0.18	7.72 ± 0.29	4.73 ± 0.06
D1_11	1.89	5.62 ± 0.09	3.54 ± 0.06	5.87 ± 0.08	7.47 ± 0.18	5.85 ± 0.07
D1_12	1.90	6.21 ± 0.12	3.38 ± 0.07	5.64 ± 0.12	7.37 ± 0.26	5.30 ± 0.06
D1_13†	1.95	5.74 ± 0.11	3.29 ± 0.07	5.52 ± 0.12	7.32 ± 0.27	3.76 ± 0.04
D1_14	2.00	7.09 ± 0.10	4.12 ± 0.05	6.35 ± 0.09	7.34 ± 0.17	4.70 ± 0.04
D1_15	2.05	6.48 ± 0.12	3.21 ± 0.07	5.19 ± 0.12	6.70 ± 0.27	3.07 ± 0.04
D1_16†	2.10	2.32 ± 0.05	2.19 ± 0.05	3.80 ± 0.09	5.97 ± 0.23	2.98 ± 0.04
D1_17†	2.15	6.04 ± 0.10	3.77 ± 0.06	5.76 ± 0.08	7.62 ± 0.18	3.43 ± 0.04
D1_18	2.20	10.5 ± 0.1	5.26 ± 0.06	8.06 ± 0.10	9.24 ± 0.16	3.70 ± 0.04
D1_19†	2.37	5.53 ± 0.08	3.60 ± 0.05	5.33 ± 0.08	5.64 ± 0.12	6.00 ± 0.07
D1_20	2.45	3.62 ± 0.05	2.96 ± 0.04	5.06 ± 0.07	6.40 ± 0.12	4.45 ± 0.05
D1_21	3.13	7.35 ± 0.09	4.17 ± 0.05	6.23 ± 0.08	8.11 ± 0.17	2.07 ± 0.02
D1_22	3.14	5.19 ± 0.07	2.94 ± 0.04	5.08 ± 0.07	5.32 ± 0.13	3.75 ± 0.05
D1_23	3.19	9.07 ± 0.13	4.57 ± 0.06	6.88 ± 0.10	7.50 ± 0.16	5.20 ± 0.06
D1_24	3.49	6.54 ± 0.09	3.69 ± 0.05	6.02 ± 0.08	7.45 ± 0.17	3.74 ± 0.04
D1_25	3.72	4.67 ± 0.06	3.13 ± 0.04	5.40 ± 0.08	6.72 ± 0.15	3.88 ± 0.05
D1_26	3.80	6.17 ± 0.08	3.48 ± 0.04	5.84 ± 0.08	7.91 ± 0.15	3.60 ± 0.04
D1_27	4.28	6.73 ± 0.09	3.54 ± 0.05	5.64 ± 0.08	6.07 ± 0.14	4.86 ± 0.06
D1_28	4.32	3.80 ± 0.05	2.69 ± 0.04	4.79 ± 0.07	6.12 ± 0.14	4.30 ± 0.05
D1_29	4.53	10.5 ± 0.1	5.56 ± 0.07	8.32 ± 0.12	8.22 ± 0.18	4.55 ± 0.05
D1_30	4.56	4.08 ± 0.05	2.95 ± 0.04	5.12 ± 0.07	5.33 ± 0.10	4.04 ± 0.05
P3_1	1.34	6.20 ± 0.08	4.75 ± 0.06	8.53 ± 0.12	9.15 ± 0.21	3.30 ± 0.04
P3_2	1.34	5.11 ± 0.06	4.63 ± 0.06	8.41 ± 0.11	9.68 ± 0.17	3.33 ± 0.04
P3_3	1.46	4.54 ± 0.07	3.47 ± 0.05	6.07 ± 0.09	7.89 ± 0.19	2.92 ± 0.04
P3_4	1.58	4.28 ± 0.06	3.47 ± 0.04	6.03 ± 0.08	7.34 ± 0.17	1.80 ± 0.02
P3_5	1.58	2.33 ± 0.03	2.90 ± 0.04	5.72 ± 0.08	7.19 ± 0.13	3.55 ± 0.04
P3_6	2.06	2.44 ± 0.04	2.44 ± 0.03	4.74 ± 0.07	5.98 ± 0.13	2.87 ± 0.03
P3_7*	2.91	2.34 ± 0.03	2.50 ± 0.03	4.96 ± 0.07	7.33 ± 0.18	2.29 ± 0.03
P3_8	3.75	8.42 ± 0.17	5.66 ± 0.12	8.65 ± 0.18	10.4 ± 0.3	2.13 ± 0.03
P3_9*	4.59	1.89 ± 0.03	2.33 ± 0.03	4.63 ± 0.07	6.67 ± 0.16	2.12 ± 0.03
P3_10*	6.28	4.11 ± 0.06	3.79 ± 0.05	6.89 ± 0.10	10.1 ± 0.2	1.58 ± 0.02
P3_11	6.76	6.18 ± 0.09	3.95 ± 0.05	6.18 ± 0.10	7.70 ± 0.19	0.86 ± 0.01
P3_12*	7.49	2.06 ± 0.03	2.58 ± 0.03	5.04 ± 0.07	7.64 ± 0.17	1.72 ± 0.02
P3_13*	9.65	2.26 ± 0.03	2.79 ± 0.04	5.38 ± 0.08	7.81 ± 0.19	2.87 ± 0.03

One key prerequisite for a successful determination of dissolved noble gas concentrations and NGTs in stalagmites is to significantly reduce the air content of a stalagmite sample before noble gas extraction from water inclusions (Kluge et al., 2008; Scheidegger et al., 2010). This is because only noble gases released from water inclusions (0.01-0.1 wt.%, Kendall and Broughton, 1978; Schwarcz and Harmon, 1976; Scheidegger et al., 2010) contain information about the temperature during the last gas exchange. Noble gases released from air inclusions (2-3 vol.%, Badertscher, 2007) lead to large excesses in the measured noble gas concentrations relative to solubility equilibrium and mask the temperature dependent noble gas signature in the water inclusions (an air content of ~ 0.5% in a sample containing 1mg of water leads to an uncertainty of ~ 10°C in the NGT calculated from Kr and Xe concentrations). Hence, prior to noble gas extraction the stalagmite samples are pre-crushed into grains of a defined diameter (350 µm) so that air inclusions, which are usually larger than water inclusions and often lie along crystal boundaries (see also Figure 5.1), are predominantly opened, whereas water inclusions are left intact. To this end, the sample is crushed in a mortar by striking the pestle with a hammer. After each hammer stroke, the crushed sample is sieved and grains < 350 µm in diameter are separated from the rest of the sample. Such "gentle" pre-crushing reduces the air content of the samples by a factor of 100 to 1000 (Scheidegger et al., 2010). Pre-crushing of stalagmite samples (4-6g) occurred in a glove box in a highly purified gas phase (He of 99.9999% purity), as under such conditions the amount of Kr and Xe adsorbing on the freshly created surfaces of the stalagmite sample was shown to be negligible (Scheidegger et al., in review).

The pre-crushed samples were stored in a stainless steel tube and connected to the noble gas extraction line without exposing the samples to air. Noble gases were finally extracted from water inclusions and the remaining air inclusions in the pre-crushed sample by heating for 1h at 320°C.

The mass of the extracted water was determined manometrically (Scheidegger et al., 2010) and the abundance of noble gases was measured by static mass spectrometry (Beyerle et al., 2000). Overall, the concentrations of Ne, Ar Kr and Xe were determined with an analytical 1σ-error of 1.5 – 3% (Table 5.1). The errors of the noble gas concentrations account for the error of the manometrically determined water mass (~ 1.5%) and the 1σ-standard deviation of the measured noble gas standard signals. NGTs were calculated from the measured Kr and Xe concentrations, which were assumed to be consistent with binary mixtures of noble gases released from air inclusions and water inclusions, i.e. air-saturated water (ASW) and atmospheric air (Scheidegger et al., in review). For this calculation we used the noble gas solubility data as recommended by Kipfer et al. (2002) and for the abundance of Kr and Xe in atmospheric air the data given in Porcelli et al. (2002). The salinity of the fluid inclusion water was assumed to be negligible (Badertscher, 2007) and the atmospheric pressure was calculated from the elevation of the cave site using the barometric formula. We note that in stalagmite D1 the binary mixing model for Kr and Xe concentrations allowed the calculation of an NGT for all 30 samples analysed. In contrast, Kr and Xe concentrations in 5 out of the 13 samples from stalagmite P3 are not consistent with such binary mixtures of ASW and atmospheric air, although the primary origin of fluid inclusions and the calcite structure in stalagmite P3 in principle suggest that Kr and Xe concentrations are unaffected by secondary processes during stalagmite growth or during noble gas analysis. No NGTs are hence presented for these samples.

Table 5.2: Results of the NGT calculation. Shown are the calculated parameters NGT and the amount of "excess air" A in cm^3 of air per cm^3 of water. Errors were calculated by Monte-Carlo error propagation of the analytical errors of Kr and Xe concentrations. The data of 7 samples from stalagmite D1 are from (Scheidegger et al., in review) and are marked with †.

Sample	Age (ka)	NGT (°C)	A (10^{-3})
D1_1	0.04	1.7 ± 1.5	133 ± 6
D1_2	0.07	56.0 ± 6.5	101 ± 4
D1_3	0.19	11.6 ± 1.6	37.1 ± 4
D1_4	0.30	25.0 ± 3.6	140 ± 6
D1_5	0.37	34.0 ± 3.3	49.9 ± 4
D1_6	0.74	34.7 ± 2.1	6.2 ± 2
D1_7	1.60	60.5 ± 2.5	16.1 ± 1
D1_8†	1.75	36.4 ± 4.0	4.4 ± 4
D1_9	1.80	25.2 ± 2.9	1.2 ± 4
D1_10†	1.85	27.9 ± 3.2	5.2 ± 4
D1_11	1.89	28.4 ± 1.8	3.5 ± 2
D1_12†	1.90	27.6 ± 2.8	0
D1_13†	1.95	27.3 ± 2.7	0
D1_14	2.00	33.9 ± 2.3	13.1 ± 2
D1_15	2.05	31.0 ± 3.0	0
D1_16†	2.10	27.6 ± 2.3	0
D1_17†	2.15	26.2 ± 1.9	0
D1_18	2.20	27.4 ± 1.7	21.6 ± 2
D1_19†	2.37	48.4 ± 2.5	14.4 ± 2
D1_20	2.45	33.3 ± 1.6	1.2 ± 2
D1_21	3.13	25.0 ± 1.6	2.8 ± 2
D1_22	3.14	50.0 ± 2.7	13.4 ± 2
D1_23	3.19	37.1 ± 2.6	20.3 ± 3
D1_24	3.49	29.7 ± 1.8	6.1 ± 2
D1_25	3.72	32.6 ± 1.8	3.5 ± 2
D1_26	3.80	24.2 ± 1.3	0
D1_27	4.28	44.7 ± 2.8	14.9 ± 2
D1_28	4.32	34.2 ± 1.8	0
D1_29	4.53	42.9 ± 3.4	37.2 ± 3
D1_30	4.56	51.4 ± 2.3	14.2 ± 2
P3_1	1.34	31.7 ± 2.5	30.8 ± 3
P3_2	1.34	25.4 ± 1.6	23.1 ± 2
P3_3	1.46	25.4 ± 1.6	27.0 ± 3
P3_4	1.58	30.5 ± 1.9	7.8 ± 2
P3_5	1.58	29.6 ± 1.5	4.1 ± 2
P3_6	2.06	35.0 ± 1.9	0.6 ± 2
P3_8	3.75	21.1 ± 2.4	19.8 ± 4
P3_11	6.76	27.9 ± 1.9	6.5 ± 3

5.4 Results and Discussion

5.4.1 Noble gas temperatures

The calculated NGTs, the water content and the amount of "excess air" for the samples we analysed in stalagmite D1 and P3 are shown in Table 5.2 and Figures 5.5 and 5.6. Results from high-latitude ice core records revealed that air temperatures remained relatively stable within 1-2 °C during the Holocene and climate variability mainly affected the hydrological cycle (Mayewsi et al., 2004; Wanner et al., 2008). Also, sea-surface temperature reconstructions in sediment cores from the Arabian Sea near Socotra Island indicate only minor temperature fluctuations of ~ 2°C between the early and the late Holocene (Jung et al., 2004). Compared to these temperature reconstructions, the calculated NGTs in stalagmite D1 and P3 reveal a much larger variation of several degrees Celsius within short periods on decadal to millennial timescales (Table 5.2 and Figures 5.5 and 5.6). In addition, some of the NGTs, particularly in stalagmite D1 are too high to reasonably reflect meteoric water temperatures in natural systems. It is hence unlikely that the NGT variation observed in the two stalagmites reflects actual fluctuations in the mean annual air temperatures on Socotra Island during the Holocene.

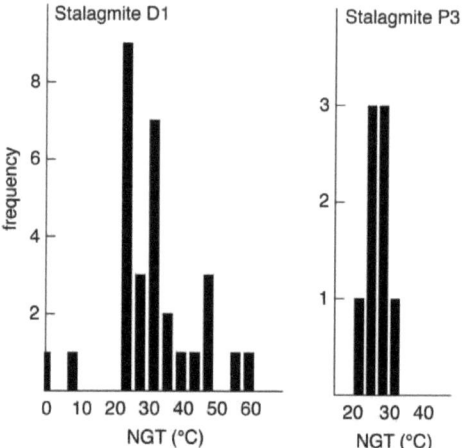

Figure 5.4: Histograms with a bin width of 4°C showing the frequency of calculated NGTs in stalagmite D1 and P3.

Figure 5.4 shows the calculated NGTs for stalagmites P3 and D1 in a histogram (bin-width of 4°C). NGTs in stalagmite P3 seem to be normally distributed with a mean value of 28.3°C and a standard deviation of 4.3°C. The mean NGT hence agrees well with the modern cave temperature. In contrast, the distribution of NGTs in stalagmite D1 seems to be skewed towards physically unreasonably high temperatures. The two samples with the lowest NGTs are regarded as outliers, as they may be affected by adsorption of residual Kr and Xe onto the sample during pre-crushing in the glove box. Discarding these two NGTs, the mean NGT in stalagmite D1 is 35.3 ± 10.2°C, a value considerably higher than the modern cave temperature. However, we note that many of the samples in stalagmite D1 (19 out of the 30 samples analysed) yield an NGT close to the modern cave temperature.

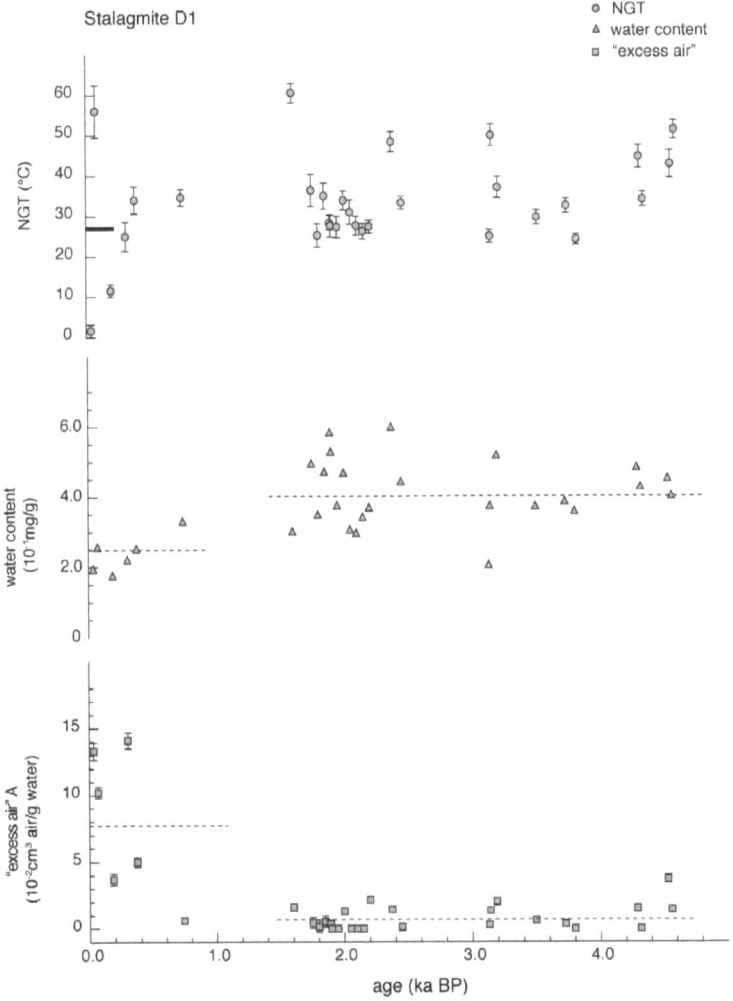

Figure 5.5: Records of the calculated NGT from Kr and Xe concentrations, the specific water content and the amount of "excess air" A in stalagmite D1. The modern mean annual air temperature of 28°C is indicated by the bold black line. The dashed line represent the mean values of the water content and the amount of "excess air" before and after the regime shift at ~ 1.6 ka BP (see text).

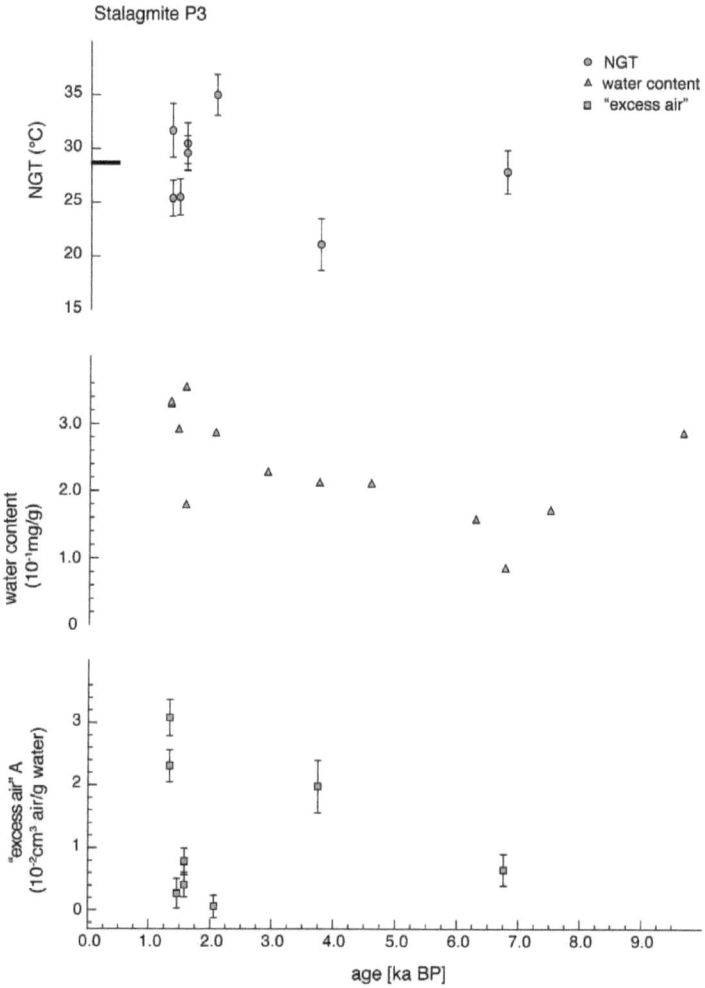

Figure 5.6: Records of the calculated NGT from Kr and Xe concentrations, the specific water content and the amount of "excess air" A in stalagmite P3. The modern mean annual air temperature of 28°C is indicated by the bold black line.

The variations in the Kr and Xe concentrations measured in stalagmites D1 and P3 are hence not attributed to temperature changes but to a different process that has the ability to affect equilibrium concentrations in the fluid inclusion water. For instance, a reduction in the partial pressures of Kr and Xe in the gas layer, where air-water partitioning occurs, would lower the equilibrium concentrations of dissolved Kr and Xe in the fluid inclusion water. Preliminary measurements of the gas extracted from air inclusions in stalagmites from poorly ventilated caves indeed revealed CO_2 contents in the air inclusions of up to 10% (Badertscher. 2007 and Chapter 6 of this thesis). Therefore, we speculate that the air layer in the vicinity of the stalagmite, where CO_2 produced during calcite precipitation seems to accumulate, may have a different gas composition than the air in the rest of the cave. If so, the absolute partial pressures of noble gases in this layer would be lowered due to the higher CO_2 partial pressure and accordingly reduce equilibrium concentrations in water inclusions. This in turn would lead to higher NGTs calculated from such concentrations.

An occasional CO_2 enriched air layer around the stalagmite with reduced partial pressures of Kr and Xe plausibly explains the bias towards high temperatures in stalagmite D1, which originates from a poorly ventilated cave. In such caves it appears likely that a gas layer with elevated CO_2 concentrations is maintained around the stalagmite. However, very high CO_2 concentrations of 50-60% would be required to yield the NGTs at the extreme end of the distribution of NGTs in stalagmite D1. Nevertheless, we suspect that the structure of the surface of a growing stalagmite is very irregular so that CO_2 partial pressures in the gas layer controlling air-water partitioning of noble gases may be highly variable on short spatial and temporal scales. Pit cave is well ventilated due to a constant airflow between the two entrances in the north and in the south of the cave, which may continuously remove the CO_2 that degasses when calcium carbonate precipitates. Therefore, in stalagmite P3 the calculated NGTs seem to be normally distributed around the modern cave temperature and are not biased towards high temperatures as in stalagmite D1. Nevertheless, the standard deviation of NGTs in Pit cave is still too large to obtain accurate and reliable cave temperature estimates, which are necessary for a quantitative interpretation of stable isotope records in stalagmites.

5.4.2 Water content and "excess air"

In Figures 5.5 and 5.6 the water content and the amount of "excess air" are shown over the time interval covered by the D1 and P3 samples. A statistical analysis of the D1 data revealed a statistically significant shift in the timeseries in the specific water content and the amount of "excess air" at ~ 1.6 ka BP. At that time the water content and the amount of "excess air" abruptly changed from higher water contents and low amounts of "excess air" before ~ 1.6 ka BP towards lower water contents and higher amounts of excess air after ~ 1.6 ka BP (Figure 5.4). This regime shift was identified by both the Pettitt-test (Pettitt, 1979), which detects the major break point in a timeseries as well as by the Rodionov-test (Rodionov, 2004) indicating statistically significant shifts in the mean value over a given timescale. The average value of the specific water content decreased from 0.41 mg/g before 1.6 ka to about half of this value (0.24 mg/g) and the amount of excess air increases by an order of magnitude after ~ 1.6 ka BP. Around the same time, the growth rate in stalagmite D1 decreased by ~ 20 % and the growth of stalagmite P3 even ceased. The concurrence of an abrupt change in the water content, the "excess air" and the growth rate in both stalagmites suggests that the conditions of stalagmite growth in the caves changed dramatically at ~ 1.6 ka BP so that less calcite was precipitated and the inclusions trapped in the calcite crystals were predominantly filled with air and less with water. The statistically significant regime shift indicates that the water content and the amount of "excess air" react to climatic changes occurring in the cave region. Therefore, these two parameters could be used as new climate proxies in stalagmites.

We note that the regime shift observed in the time series of the water content and in the amount of "excess air" roughly coincides with the major break point in the $\delta^{18}O_{calcite}$ record of stalagmite D1, at 1.3 ka BP, as determined by the Pettitt-test (Figure 5.7). In stalagmite P3, a minimum in $\delta^{18}O_{calcite}$ seems also to be present at 1.3 ka BP, although it does not stand out as clearly as in stalagmite D1 (Figure 5.7). Hence, the concurrence of abrupt changes in different time series seems likely to indicate that the change in growth conditions expressed by the water content and the amount of "excess air" is linked to the observed minimum in the $\delta^{18}O_{calcite}$ records of stalagmite D1 and P3.

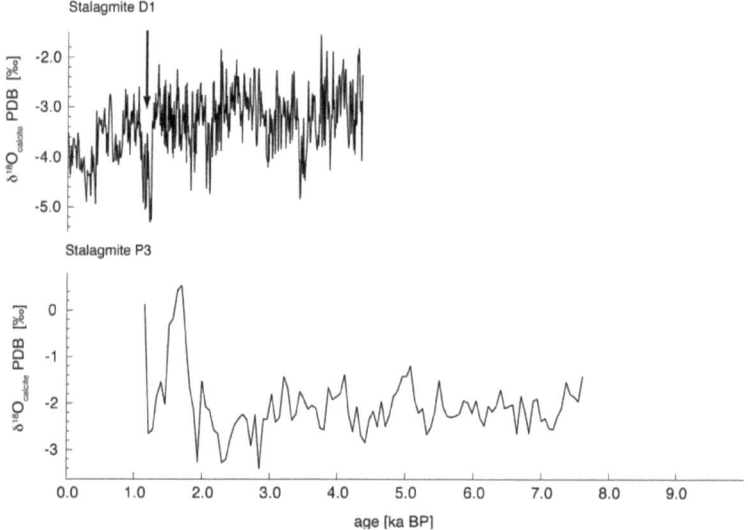

Figure 5.7: Oxygen isotope records ($\delta^{18}O_{calcite}$) of stalagmite P3 and D1 (Fleitmann et al., 2007 for stalagmite D1, unpublished data for stalagmite P3). The arrow indicates the major break point in the D1 timeseries (Pettitt, 1979).

These changes may also be linked to the first (~ 1.5 ka BP) of several events of abrupt climate change during the Holocene, which were observed in North

Atlantic sediment cores and in the data of other climate proxies from the Northern Hemisphere (Bond et al., 1997; 2001; Mayewski et al., 2004). This event may hence be captured via teleconnections in the stable isotope record as well as in the amount of air and water inclusions being formed in the stalagmites on Socotra Island.

Previous work on stalagmites from Southern Oman and Socotra Island has shown that $\delta^{18}O_{calcite}$ mainly records changes in the amount of precipitation (Fleitmann et al., 2007; Burns et al., 2001; 2003; Shakun et al., 2007), with more negative values corresponding to greater rainfall and more positive values being associated with decreased precipitation ("amount effect" see Dansgaard, 1964; Rozanski et al., 1993). According to Fleitmann et al. (2007), the modest long-term decrease in $\delta^{18}O_{calcite}$ which is observed in both stalagmites, indicates an increase in the amount of precipitation over the Holocene, maybe due to an earlier retreat of the ITCZ from its northernmost position and accordingly an earlier onset and also a longer duration of the NE-Monsoon (Fleitmann et al., 2007). Such increased precipitation may also be reflected in the increasing specific water content since 7 ka BP in stalagmite P3 (Figure 5.5). We speculate, that the return to drier conditions at the end of the $\delta^{18}O_{calcite}$ minimum induced the "regime shift" observed in the timeseries of the water content and the amount of "excess air" in D1 and the end of stalagmite growth in Pit cave. Stalagmite growth rates are mainly a function of the cave temperature, the calcium concentration of drip waters, of drip water availability and of the partial pressure difference between the soil air CO_2 and cave air CO_2 (McDermott et al., 2005). Therefore, growth conditions may have changed around ~ 1.6 ka towards a reduced drip rate, leading to a lower abundance of water inclusions and to a reduced stalagmite growth rate. Accordingly, air inclusions were formed more frequently, presumably because the surface of the stalagmite was more likely to be exposed to cave air. Also, calcium concentrations may have been lowered in response to reduced biological activity in the soil above the cave during such dry periods. These reduced growth conditions seem to have prevailed until today.

5.5 Conclusions and Outlook

In this study, precise Kr and Xe concentrations were determined in fluid inclusions in samples from two Holocene stalagmites from Socotra Island. For the majority of samples, Kr and Xe concentrations are consistent with a two-component mixture of ASW and atmospheric air, allowing the determination of an NGT. However, the large variation in NGTs observed in both stalagmites seems more likely to be linked to changing conditions during air-water partitioning than to actual changes in the cave temperature. We hypothesize that CO_2, which degasses during calcite precipitation, may accumulate in the air layer around the stalagmite and reduce the partial pressures of noble gases accordingly. To verify this hypothesis, direct analysis of the composition of the air layer around a stalagmite should be conducted in the future in order to determine the real partial pressures of noble gases during air-water partitioning in the close proximity of the growing stalagmite surface. We conclude that the knowledge of the local conditions prevailing during the gas exchange process of noble gases between the CO_2 enriched air layer around the stalagmite and the fluid inclusion water seem to be crucial for obtaining accurate and precise NGTs in stalagmites. The gas inclusion process hence needs to be understood in much more detail.

In our study we also identified the timeseries of the specific water content and the amount of "excess air" as new potential climate proxies in stalagmites. Both parameters seem to contain useful information about the growing conditions during calcite deposition and stalagmite growth. Therefore, the water content and the amount of "excess air" offer the possibility to contribute with additional climatic information to the interpretation of oxygen isotope records in stalagmites.

Acknowledgements

The authors would like to thank Urs Menet and Heinrich Baur for their assistance in the laboratory.

6

Trace gas analysis in air inclusions of stalagmites

Abstract During the growth of stalagmites, drip water as well as cave air are incorporated as fluid inclusions into the growing calcite crystals. So far, scientific studies have focused only on the analysis of water inclusions. However, the existence of air inclusions offers the possibility to study the gases in air inclusions as a potential proxy of the past composition of the surrounding cave air. The aim of this pilot study is to assess the potential of air inclusions and their atmospheric trace gas composition as a new climate and environmental proxy in stalagmites. We analysed samples from two different stalagmites, which are known to contain air inclusions and where the isotope composition of the calcite is known. The measurement of the major elemental composition (Ar, N_2, O_2) as well as the isotope composition of trace gases (CO_2 and CH_4) indicates that the gases in air inclusions indeed carry some information about the local environment in the cave.

6.1 Introduction

Interest in fluid inclusions in stalagmites as proxies for past environmental conditions such as cave temperatures or the isotope composition of drip waters was so far focused on water inclusions only (van Breukelen et al., 2008; Fleitmann et al., 2003b; Zhang et al., 2008, Kluge et al., 2008; Scheidegger et al., 2010) as the presence of air inclusions was long unknown (Badertscher et al., 2007; Kluge et al., 2008). Air inclusions (10-200 µm) in stalagmites make up to 2-3% of the stalagmite volume and are either located between or within calcite crystals (inter- vs. intra-crystalline). First systematic microscopic analyses found that air inclusions are clearly physically separated from water inclusions. Also, no evidence was found that air inclusions might be part of open pores connected with the surface of the stalagmite (Badertscher, 2007; Scheidegger et al., 2007a). Hence, in principle air inclusions have the potential to be used as climate archives similar to air that is trapped in bubbles in ice sheets (e.g. Jouzel et al., 2007; Petit et al., 1999; Andersen et al., 2004). However, in contrast to air inclusions in ice, which are "sub-samples" of free atmospheric air, air inclusions in stalagmites are filled with cave air from the proximity of the growing stalagmite, which does not necessarily have the same composition as free air.

Generally, the process of stalagmite growth, the composition of the air in the soil above the cave, the cave volume and the rate of cave ventilation are the main factors determining the composition of cave air (Kowalczk and Froelich, 2010). Hence, the composition of cave air is, particularly in the case of trace gases, very site specific and it is expected that it differs from the composition of the free atmosphere. Especially CO_2 is enriched in caves, as it is actively formed in response to stalagmite deposition and emanates to the cave air (e.g. Spötl et al., 2005; Baker and Genty, 1998; Kowalczk and Froelich, 2010). Other trace gases (e.g. CH_4, CO_2 and N_2O) are likely to enter the cave directly through fissures from the soil air above the cave. Hence, trace gases in air inclusions could potentially provide information about e.g. the growth rate of stalagmites and the biological activity in the soil above the cave. In addition, the isotope composition of trace gases in air inclusions ($\delta^{13}C_{CO2}$, $\delta^{13}C_{CH4}$, $\delta^{18}O_{CO2}$) could give insights into geochemical origin of the respective trace gases.

We present preliminary results of the first pilot analyses of the abundance of major atmospheric gases as well as the isotope composition of the trace gases CO_2 and CH_4 in the gas extracted from air inclusions in stalagmites. Our results show that cave air is indeed trapped in the stalagmite at the time of $CaCO_3$ deposition and stored over the long timescales of stalagmite growth. Also, the oxygen and carbon isotope composition of the liberated CO_2 from air inclusions seems to be linked to the isotope composition of the stalagmite $CaCO_3$. The observed relation may indicate that information about the climatic and environmental conditions is preserved in the gas composition of air inclusions, although an exact and final interpretation of the results is not yet possible. Nevertheless, our results are very promising and pave the way to investigate this emerging climate archive in stalagmites in more detail in the future.

6.2 Material and Methods

6.2.1 Samples

We analyzed 9 samples covering the last 1.5 ky from stalagmite Y4 (Yenesu cave, Turkey) and 9 samples from stalagmite SO2 (Sofular cave, Turkey), whose growth history goes back to the last glacial maximum (LGM). Both stalagmites were collected in parts of the cave, where the relative humidity reaches 100%. We hence assume that the stalagmites were deposited in isotope equilibrium, as no kinetic fractionation due to evaporative processes occurs at such humidity (Hendy, 1971). U/Th-dating of the samples as well as the analysis of the stable isotope composition of the $CaCO_3$ ($\delta^{18}O$ and $\delta^{13}C$) were carried out at the Geological Institute in Bern (Fleitmann et al., 2009 for stalagmite SO2 and unpublished data for Y4).

6.2.2 Gas analysis

The abundance of major atmospheric gases in air inclusions and their isotope composition were analysed at the Physics Institute of the University of Bern. Stalagmite samples were cut from the centre of the stalagmite into pieces of ca. 5 x 0.5 x 0.5 cm^3 (corresponding to 1-2 g, temporal resolution 50-200 years) to fit into the copper tubes we used as sample containers. To avoid contamination (e.g. filling of "open" air inclusions with atmospheric air or adsorption of atmospheric gases on the surface), the samples were stored under vacuum and heated to 50°C until they were analyzed. To extract the gases from air inclusions under vacuum conditions, we squeezed the copper tube containing the sample using a hydraulic hand pump, which pressed a stainless steel plate onto the copper tube, and hence allowed to crush the sample. Such gentle crushing breaks the calcite predominantly along crystal boundaries and mainly opens inter-crystalline air inclusions (Scheidegger et al., 2010). As crushing may also open some water inclusions, the liberated gas first passed a cold trap held at a temperature of -80°C to freeze out the liberated water.

The abundance of major atmospheric gases (Ar, N_2, O_2, CO_2) in air inclusions was determined by mass spectrometry using standard analytical protocols for the analysis of air samples taken from e.g. ice cores (see Leuenberger et al., 2000a; Leuenberger et al., 2000b). Compared to the gas amounts that are usually measured with this protocol, the gas amounts extracted from stalagmite samples are often close to the detection limit of the applied method. As a consequence, these first results of the analysis of major atmospheric gases can only be interpreted semi-quantitatively.

In contrast, the oxygen and carbon isotope composition of the liberated trace gases CO_2 and CH_4 were precisely determined by GC/MS, using an analytical protocol especially developed for carbon and oxygen isotope analysis of very small air amounts (Leuenberger et al., 2003). Prior to gas extraction by crushing, the composition of the gas, which seemed to emanate from the bulk sample, was continuously analysed until the amplitude on m/z 44 (CO_2) reached the blank level. Crushing the stalagmite sample then released CO_2 and CH_4 amounts similar to those of air samples usually measured with the same analytical protocol. The isotope composition of CO_2 and CH_4 could therefore be determined with an overall analytical 1σ-error of 0.1 ‰.

6.3 Results and Discussion

6.3.1 Major elemental composition

The results of our qualitative measurements of the major atmospheric gases in the air inclusions of the two stalagmites are presented in Table 6.1 and Figure 6.1. The results are reported as the relative difference of O_2/N_2, Ar/N_2 and CO_2/N_2 against the respective ratios in atmospheric air.

$$\delta\left(\frac{i}{N_2}\right) = \left(\frac{\left(\frac{i}{N_2}\right)_{sample}}{\left(\frac{i}{N_2}\right)_{air}} - 1\right) \times 1000 \quad (‰)$$

The results reveal that air inclusions contain O_2, N_2 and Ar in near atmospheric abundance, as the relative deviation of each gas from the composition of free air is in maximum only a few percent. In contrast, CO_2 is strongly enriched relative to its content in free air (up to two orders of magnitude). This indicates that air inclusions contain trapped cave air, as cave air is known to have elevated CO_2 levels.

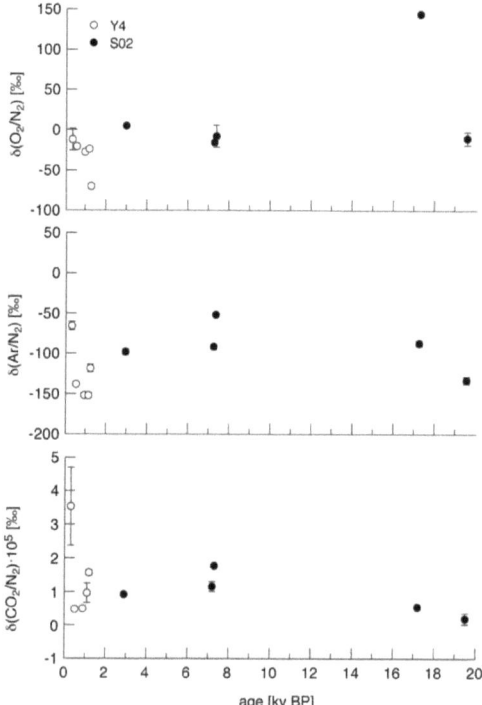

Figure 6.1: Relative deviation of the abundance of major atmospheric gases (O_2, N_2, Ar and CO_2) measured in the gas extracted from 5 samples of stalagmite Y4 and SO2 respectively in comparison to their abundance in free air.

Using the data in Table 6.1, we determined the CO_2 content in air inclusions (Table 6.1), assuming that the N_2 abundance in cave air and air inclusions is the same as in the free atmosphere, and using the pre-industrial atmospheric CO_2 content of 0.028 vol.%. Compared to other studies, which analysed cave air by direct instrumental measurements in the cave environment (Kowalczk and Froelich, 2010; Spötl et al., 2005; Banner et al., 2007; Baldini et al., 2008; Baldini, 2010), our estimated CO_2 contents are 10 to 100 times larger. Such high CO_2 contents are unlikely to reflect the cave CO_2 content in general and may hence rather reflect the CO_2 level in the proximity of the growing stalagmite surface.

Table 6.1: Measured ratios of the major atmospheric gases in respect to atmospheric air, and the calculated CO_2 abundance ($CO_{2,calc}$), assuming an atmospheric N_2 abundance in the cave air. The errors are denoted as the 1s-errors of the 5 analyses done in each measurement cycle.

Sample	Age (ka)	$\delta(O_2/N_2)$ (‰)	$\delta(Ar/N_2)$ (‰)	$\delta(CO_2/N_2)$ (10^4‰)	$CO_{2,est}$ (vol.%)
Y4_1	0.3	-12.0 ± 13.3	-65.5 ± 5.1	35 ± 12	9.4
Y4_2	0.5	-20.6 ± 0.1	-138.1 ± 0.1	4.8 ± 0.3	1.4
Y4_3	0.9	-27.6 ± 0.1	-152.1 ± 0.1	4.9 ± 0.3	1.4
Y4_4	1.1	-23.9 ± 0.9	-152.2 ± 3.3	9.6 ± 2.9	2.7
Y4_5	1.2	-69.8 ± 2.7	-118.6 ± 4.6	16 ± 0.8	4.4
SO2_1	2.9	4.9 ± 1.2	-98.1 ± 3.5	9.2 ± 0.5	2.6
SO2_2	7.2	-15.6 ± 2.2	-91.7 ± 2.9	12 ± 1.4	3.3
SO2_3	7.3	-7.8 ± 13.7	-51.6 ± 1.3	18 ± 0.4	5.0
SO2_4	17.2	143.9 ± 1.7	-87.1 ± 3.5	5.5 ± 0.2	1.6
SO2_5	19.5	-10.8 ± 8.1	-133.3 ± 4.5	2.0 ± 1.6	0.6

Therefore, we speculate that the elevated CO_2 abundance in air inclusions indicates the existence of a CO_2 enriched air layer around the stalagmite, which has a different gas composition than the rest of the cave. This gas layer may continuously be fed by CO_2 that degasses during calcite precipitation. As a result, the local CO_2 abundance near the stalagmite surface may be much higher than it is in the rest of the cave. We hence propose that a CO_2 concentration gradient is maintained between the surface of the stalagmite through the gas boundary layer towards the open cave air. Then, the rate of calcite precipitation, i.e. the growth rate of a stalagmite is not only determined by the diffusion of CO_2 through the water film on the surface of the stalagmite (Dreybrodt, 1980) but also by renewal of the CO_2 enriched air layer around the stalagmite. The CO_2 abundance in air inclusions may then potentially provide information about the stalagmite growth rate. This could serve as a potential explanation for the lower CO_2 contents measured in the SO2 samples from the LGM (SO2_4 and SO2_5), where growth rates were smaller than today due to drier conditions (data provided by D. Fleitmann, University of Berne).

6.3.2 Stable isotope composition of CO_2 and CH_4

To better understand the geochemical origin of the measured CO_2 in the air inclusions and to get a more precise interpretation of their trace gas content, we discuss in this section the results of stable isotope analysis of CO_2 and CH_4 (Table 6.2). We note that measurable amounts of CH_4 were found to be present in air inclusions, indicating elevated CH_4 concentrations compared to the free atmosphere. However, a quantification of CH_4 concentrations in air inclusions is not yet possible.

Table 6.2: The isotope composition of CO_2 and CH_4 in the gas extracted from air inclusions after crushing, and the calculated isotope composition of CO_2 in equilibrium with $CaCO_3$. Also shown is the amplitude on m/z 44 (CO_2) as an indication of the amount of extracted gas (blank level 30 to 100 mV). 5 samples were cut into two aliquots before the analysis and measured separately. They are indicated with (1) and (2).

Sample	Age (ka)	Mass44 (mV)	$\delta^{13}C$ (‰ PDB)		$\delta^{18}O$ (‰ PDB)	$CO_{2, calculated}$ (‰ PDB)	
			CO_2	CH_4	CO_2	$\delta^{13}C$	$\delta^{18}O$
Y4_6(1)	0.2	2168	-43.6	-30.7	0.5	-21	4.9
Y4_6(2)	0.2	4291	-42.9	-19.8	0.3		
Y4_7	0.4	682	-42.8	-28.1	6.8	-21	4.9
Y4_8	0.6	2039	-42.4	-21.1	5.6		
Y4_9(1)	1.0	2468	-37.2	-21.9	5.6	-21	4.9
Y4_9(2)	1.0	2974	-60.5	-36.7	-0.9		
SO2_6	3.4	1850	-54.6	-20.7	-1.1	-20	4.2
SO2_7(1)	5.0	2172	-45.3	-36.7	3.5	-20	4.0
SO2_7(2)	5.0	2912	-50.8	-23.6	1.7		
SO2_8(1)	7.3	6899	-48.1	-22.7	-0.9	-20	3.3
SO2_8(2)	7.3	4068	-43.9	-26.5	1.0		
SO2_9(1)	19.5	1341	-36.3	-28.5	-1.9	-17	-0.5
SO2_9(2)	19.5	1064	-32.9	-31.6	-5.9		

Figures 6.2 and 6.3 show the $\delta^{18}O$ of CO_2 and $\delta^{13}C$ of CO_2 and CH_4 released from the air inclusions in comparison with the respective isotope composition of the calcite. In both stalagmites $\delta^{13}C$ of CO_2 is lower than $\delta^{13}C$ of the calcite ($\delta^{13}C_{calcite}$). The opposite is true for $\delta^{18}O$ of CO_2 in air inclusions and calcite ($\delta^{18}O_{calcite}$), respectively. This isotopic shift agrees with the known isotope fractionation during calcite deposition, stating that calcite precipitated in isotope equilibrium with CO_2 gas is enriched in ^{13}C and depleted in ^{18}O (Romanek et al., 1992; Kim and O'Neil, 1997; O'Neil and Adami, 1969). In stalagmite SO2 the isotope composition of CO_2 liberated from air inclusions seems to follow the trend observed in $\delta^{18}O$ and $\delta^{13}C$ of the calcite over the time period of 20 ka. These results indicate that the isotope composition of trace gases in air inclusions potentially provides information about the climatic conditions at the time of incorporation in the growing stalagmite. In order to further constrain whether the CO_2 in the air inclusion carries the same climatic information as the calcite, i.e. if the precipitated $CaCO_3$ and the CO_2 in the air inclusions are in isotope equilibrium, we calculated $\delta^{13}C$ and $\delta^{18}O$ of CO_2 in equilibrium with $CaCO_3$ using the fractionation factors and the cave temperatures given in Table 6.3.

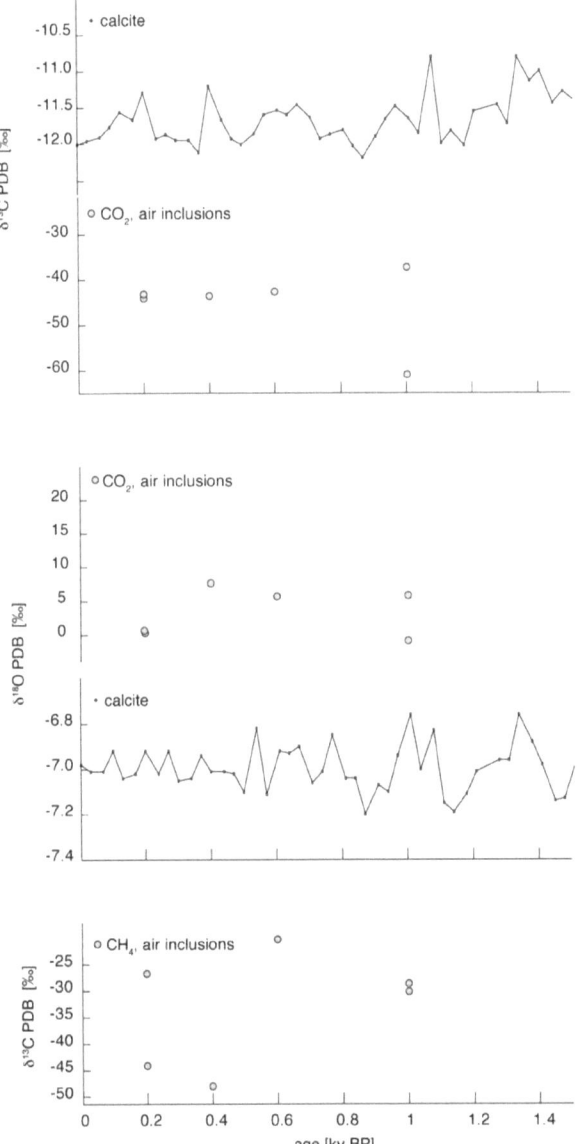

Figure 6.2: The stable isotope composition of CO_2 and CH_4 measured in the gas extracted from stalagmite Y4 and the stable isotope composition of the calcite (data from D. Fleitmann, unpublished).

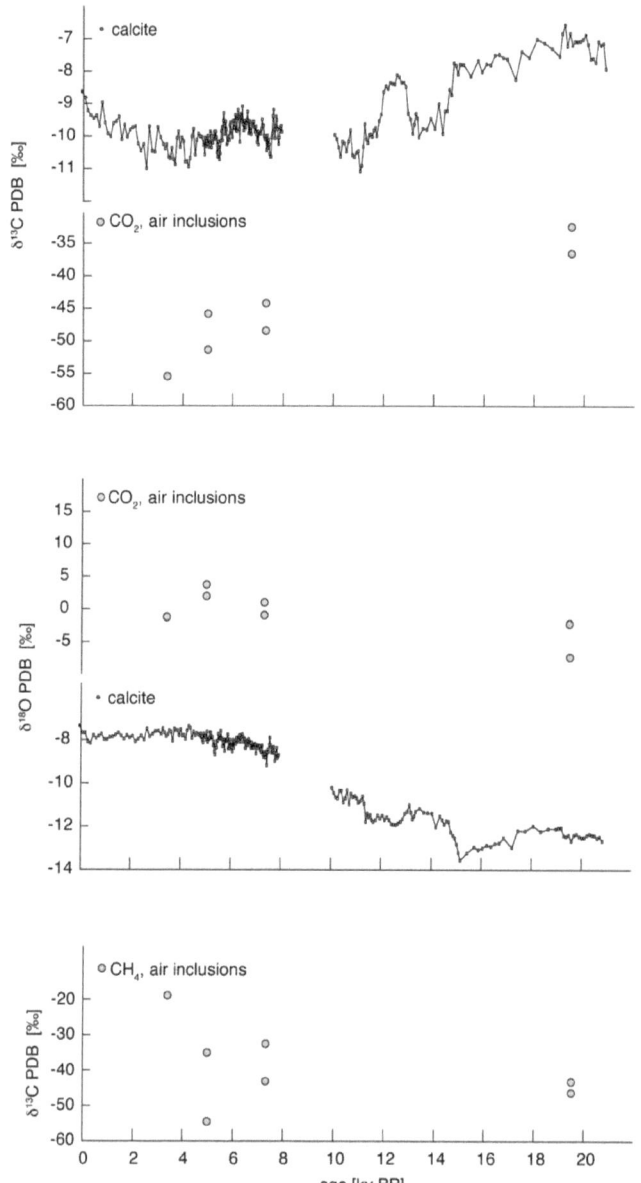

Figure 6.3: The stable isotope composition of CO$_2$ and CH$_4$ measured in the gas extracted from stalagmite SO2 and the stable isotope composition of the calcite (data from Fleitmann et al., 2009).

Comparison of the calculated values with the values measured in the air inclusions shows that the $\delta^{18}O$ of the CO_2 in air inclusions is closer to equilibrium (deviations 0.5-5‰) than the $\delta^{13}C$, which strongly deviates from equilibrium conditions (by 20-30‰). This depletion may indicate that the CO_2 in air inclusions is a mixture of several CO_2 components, that have different carbon sources, but probably the same oxygen source. Assuming that the stalagmites were precipitated in isotope equilibrium, the first component must have a carbon isotope composition of -17 to -21‰, corresponding to the calculated equilibrium composition of degassing CO_2. The measured "mixed" $\delta^{13}C$ of -40 to -60‰ in CO_2 of air inclusions can hence only be explained by the addition of a very light carbon source. This excludes e.g. atmospheric CO_2 ($\delta^{13}C$ of -7‰).

Table 6.3: Fractionation factors a used to calculate the isotope composition of CO_2 in equilibrium with $CaCO_3$, and of CH_4 in equilibrium with CO_2 for 12° (present cave temperature) and for the assumed temperature of 8°C in Sofular cave prior to the LGM (personnel communication D. Fleitmann, University of Berne, 2009).

System	T (°C)	α (-)	References
Carbon			
calcite-CO_2	13°C	1.0104	Romanek et al., 1992
	8°C	1.0110	
CO_2-CH_4	13°C	1.0751	Battinga, 1969
	8°C	1.0772	
Oxygen			
CO_2-calcite	13°C	1.0119	Kim and O'Neil, 1979;
	8°C	1.0118	O'Neil and Adami, 1969

Only very light biogenic CH_4, with a $\delta^{13}C$ of -130 to -110‰, oxidized to CO_2, could potentially lead to the light $\delta^{13}C$ values in CO_2 of the air inclusions (Bottinga, 1969, Table 6.3). However, to our knowledge no such light CH_4 is known and the $\delta^{13}C$ of the CH_4 extracted from the air inclusions lies between -36 and -20‰. As a result, a methanogenic origin of the second CO_2 component thus seems unlikely.

As a result of these findings we conclude that the fractionation processes leading to the measured isotope composition of CO_2 and CH_4 in the gas extracted from air inclusions do not only reflect climatic effects. Additional non-climatic fractionation processes seem to affect the isotope composition of the gases trapped in air inclusions to such an extent, which up to know compromises a simple and straightforward interpretation of the isotopic signals in terms of environmental conditions. Potential non-climatic fractionation processes are e.g. diffusion of gases from air inclusions into the calcite crystal lattice after the formation of air inclusions, adsorption of gases onto the walls of air inclusions or exchange of trace gas C and O isotopes with the same isotopes of the surrounding calcite (Schwarcz et al., 1976; Dennis et al., 2001).

6.4 Conclusions and Outlook

For the first time we have analysed the gases trapped in air inclusions in stalagmites and measured the major atmospheric components as well as the isotope composition of CO_2 and CH_4. Our results reveal that trace gases such as CO_2 and CH_4 are enriched in the cave atmosphere, while the major gas composition in air inclusions is similar to the composition of free air. We also found that a climate signal may be preserved in the isotope composition of CO_2 as it seems to follow the isotope composition of the surrounding calcite over long timescales. This indicates that air inclusions may represent an archive of cave air and have the potential to be used as a new proxy in stalagmites.

However, whereas the oxygen isotope composition of the CO_2 reasonably corresponds to the values expected for equilibrium with the isotope composition of the calcite, the $\delta^{13}C$ values of the CO_2 are much more negative than expected for isotope equilibrium. Based on the data that is currently available, the identification of the responsible processes for the observed depletion remains speculative and needs to be further studied in the future. Most important for future analysis is to analyse cave air samples from different locations within the cave and compare the results with the analysis of air inclusions in modern stalagmite samples. This would help to answer the question whether the composition of the gas trapped in air inclusions is representative for the cave air in general. Also, it would potentially help to identify and disentangle the processes, which influence the isotope composition of trace gases in air inclusions. Further, the analytical protocol should be expanded to allow also the precise determination of CO_2 and CH_4 amounts. Such a combination of quantitative gas measurements and isotope analyses may then allow to study the evolution of cave air composition in a quantitative way over geological timescales and allow a more precise and detailed interpretation.

Acknowledegements

Thanks go to Markus Leuenberger and Peter Nyffeler (Climate and Environmental Physics, University of Bern) for providing the facilities in the lab for stalagmite samples and for their assistance during the analysis. Thanks also go to Seraina Badertscher, who microscopically analysed air and water inclusions in stalagmites in her master thesis. The results encouraged us to conduct this pilot study.

7

Synthesis and Outlook

The idea of applying the well-established concept of noble gas temperatures (NGTs), which is successfully employed in lakes and groundwater, to fluid inclusions in stalagmites emerged quite some years ago (e.g. Ayliffe et al., 1993). However, it was often discarded after some preliminary analysis due to analytical challenges and due to difficulties in the interpretation of the measured noble gas concentrations in terms of NGTs (Ayliffe et al., 1993; McDermott et al., 2005). The development of analytical methods to overcome the analytical limitations observed in the past was the first target of this PhD thesis, which was successfully completed. We are now able to precisely determine noble gas concentrations in water masses as small in fluid inclusions in stalagmites. The developed analytical protocol can in principle be applied to any other solid sample containing small amounts of water and where noble gas concentrations may be of interest. This could for instance include mussel shells and the fossil parts of corals to reconstruct lake and ocean water temperatures. Also, in consolidated sediments, pore water analysis could give insights into e.g. transport processes and their evolution within the sediments.

In this thesis, the developed analytical protocol was applied to stalagmite samples from several caves located in different climatic zones with different mean annual air temperatures. The goal was to precisely determine noble gas concentrations and to convert them into NGTs. Our results demonstrate that stalagmites are indeed suitable for NGT determination, as NGTs calculated from Kr and Xe concentrations in modern stalagmite samples agree well with modern cave temperatures. However, NGTs in stalagmites cannot be determined using the standard least squares fitting models commonly used in groundwater studies, as these models require a two-component mixture of air and ASW for all noble gases. As shown in this thesis, in stalagmites, the concentrations of He, Ne and Ar cannot be interpreted as simple binary mixtures of air and ASW, as they are affected by additional noble gas components (lattice trapped He and Ne and adsorbed Ar). Therefore, only the concentrations of Kr and Xe are used for NGT determination in stalagmites, as these two concentrations are not affected by the above mentioned additional noble gas components and are solely composed of noble gases from ASW and atmospheric air.

Our results further indicate that not all stalagmites are equally suitable for NGT determination. Kr and Xe concentrations in some samples seemed to have been altered either during stalagmite growth or the analytical procedure to determine noble gas concentrations so that the temperature information is no longer preserved in such samples. Our results give strong indication that Kr and Xe concentrations are well preserved in fast growing stalagmites with relatively small and loosely arranged calcite crystals, that have a high abundance of intra-crystalline water inclusions. Sample pre-selection using light as well as electron microscopes to study the structure and the spatial arrangement of calcite crystals and the origin of air and water inclusions hence is a key factor for a successful and accurate NGT determination.

The thesis further showed that the gas inclusion process of noble gases into the fluid inclusion water might be more complicated than initially expected. Our results indicate that partial pressures of noble gases near the growing stalagmite might occasionally be lower than in open cave air due to the existence of a CO_2 enriched air layer around the stalagmite, which reduces the equilibrium concentrations of Kr and Xe in the forming water inclusions. The relatively large variation of NGTs observed in Holocene samples from stalagmite D1 from Socotra Island were hence conceptually explained by variable CO_2 abundances during gas exchange and not to actual variations in the mean annual air temperature.

Overall, this thesis has paved the way for a successful application of the "noble gas thermometer" to fluid inclusions in stalagmites. The ultimate goal for the future is to use accurately and reliably determined NGTs to obtain a temperature-corrected $\delta^{18}O$ record of the calcite. To achieve this goal, the following tasks are seen as key aspects in future research on noble gas analysis in fluid inclusions in stalagmites:

Task A: Improvements of analytical methods

Although this thesis strongly focused on the development of analytical methods to determine noble gas concentrations and NGTs in stalagmites there are still some analytical aspects that need to be improved. For instance, if the adsorption of Ar during pre-crushing of stalagmite samples in the glove box could be totally avoided, Ar concentrations could be included in the calculation of NGTs. As Ar is less soluble in water than Kr and Xe, it is more sensitive to the amount "excess air", allowing the noble gases released from air inclusions to be more precisely parameterised. Hence, NGTs could be determined also in stalagmite samples, where the separation of air inclusions from water inclusions by pre-crushing is difficult, for instance because air and water inclusions have similar sizes. The use of three noble gas concentrations may also allow determining three unknown parameters, i.e. the temperature, the amount of "excess air" and the atmospheric pressure. This would allow determination of the true partial pressures of noble gases in the CO_2 enriched air layer around the stalagmite and eventually lead to the determination of accurate NGTs also for samples with high and strongly varying CO_2 concentrations in the gas layer around the stalagmite. The prevention of Ar adsorption during pre-crushing may be achieved by further reducing the residual Ar content in the gas phase of the glove box. Alternatively, a new crusher could be developed, that allows crushing the sample in vacuum into grains of a defined diameter before noble gas extraction from the pre-crushed sample by heating.

Another open question is whether Kr and Xe concentrations, that were not consistent with binary mixtures of ASW and atmospheric air, are altered during the analytical procedure to determine noble gas concentrations or if they reflect the initial concentrations trapped in the inclusions during stalagmite growth. This question could be addressed by conducting adequate experiments to assess if e.g. diffusional Kr and Xe loss occurs due to the artificially applied large concentration gradient during pre-crushing of the sample in the virtually Kr and Xe free gas phase in the glove box or during the connection of the sample to the ultra-high vacuum system.

Task B: Petrographic studies of stalagmites
The obtained results of this thesis indicate that the origin of fluid inclusions and the arrangement of the individual calcite crystals within a stalagmite seem to be a crucial factor for a successful interpretation of the measured noble gas concentrations in terms of NGTs. It is hence important to study the characteristics of calcite crystals and their fluid inclusions in selected stalagmites first of all by microscopic examinations (light and electron microscopy) and if possible also using other imaging methods (e.g. X-ray tomography and NMR). This may help to identify the most suitable stalagmite samples for NGT determination. Also, it may give some indication about how secondary transformation processes (e.g. crystal coalescence, gas and pore water migration) might affect the fluid inclusions and their noble gas content. Combining the results of petrographic examinations with records of climate proxies measured in the same stalagmite might also allow recognition of the role climate parameters play in the formation of the different types of calcite in stalagmites.

Task C: Case studies
Future work should also focus on the application of the developed tools to stalagmite samples with the goal to determine past cave temperatures on local scales and over longer timescales than those investigated within this thesis. This could for instance include the transition from the last glacial period into the Holocene as well as the temperature regime during the Eemian period (~ 120 ka BP). The specific water content, the amount of "excess air" and trace gas analysis in air inclusions could be used as complementary proxies within such climate studies. For the complementation of this task, suitable samples need to be selected based on microscopic analyses. This may constitute a

crucial point in the implementation of this task. If robust NGTs could be determined, they could then be used to partition the effect of the cave temperature on the isotope composition of the calcite ($\delta^{18}O$, $\delta^{13}C$) from other climate driven parameters. Also, a sound monitoring of the modern climate system is necessary to reduce uncertainties related with the interpretation of the calculated $d^{18}O_{dripwater}$ records even if quantitative and accurate cave temperature estimates from noble gas concentrations are available

Task D: Comparison of temperature proxies in stalagmites
Much effort is currently put in the development and the application of new methods to quantitatively determine cave temperatures in stalagmites (e.g. oxygen isotope fractionation between calcite and water, clumped isotope thermometry, noble gas concentrations in fluid inclusions). All these methods are relatively new and hence still suffer from some analytical and archive-specific shortcomings. A comparison and cross calibration of the various temperature proxies applied to the same stalagmite samples would offer the possibility to identify specific factors, which constrain the determination of the cave temperature of each temperature proxy and potentially help to improve each of the temperature methods involved.

References

Aeschbach-Hertig, W., Hofer, M., Schmid, M., Kipfer, R., and Imboden, D.M., (2002). The physical structure and dynamics of a deep, meromictic crater lake (Lac Pavin, France). *Hydrobiologia*, v. 487, p. 111-136.

Aeschbach-Hertig, W., Peeters, F., Beyerle, U., and Kipfer, R., (1999). Interpretation of dissolved atmospheric noble gases in natural waters. *Water Resources Research*, v. 35, p. 2779-2792.

Aeschbach-Hertig, W., Peeters, F., Beyerle, U., and Kipfer, R., (2000). Palaeotemperature reconstruction from noble gases in ground water taking into account equilibration with entrapped air. *Nature*, v. 405, p. 1040-1044.

Affek, H.P., Bar-Matthews, M., Ayalon, A., Matthews, A., and Eiler, J.M., (2008). Glacial/interglacial temperature variations in Soreq cave speleothems as recorded by clumped isotope thermometry. *Geochimica Et Cosmochimica Acta*, v. 72, p. 5351-5360.

Andersen, K.K., Azuma, N., Barnola, J.M., Bigler, M., Biscaye, P., Caillon, N., Chappellaz, J., Clausen, H.B., DahlJensen, D., Fischer, H., Fluckiger, J., Fritzsche, D., Fujii, Y., Goto-Azuma, K., Gronvold, K., Gundestrup, N.S., Hansson, M., Huber, C., Hvidberg, C.S., Johnsen, S.J., Jonsell, U., Jouzel, J., Kipfstuhl, S., Landais, A., Leuenberger, M., Lorrain, R., Masson-Delmotte, V., Miller, H., Motoyama, H., Narita, H., Popp, T., Rasmussen, S.O., Raynaud, D., Rothlisberger, R., Ruth, U., Samyn, D., Schwander, J., Shoji, H., Siggard-Andersen, M.L., Steffensen, J.P., Stocker, T., Sveinbjornsdottir, A.E., Svensson, A., Takata, M., Tison, J.L., Thorsteinsson, T., Watanabe, O., Wilhelms, F., and White, J.W.C., (2004). High-resolution record of Northern Hemisphere climate extending into the last interglacial period. *Nature*, v. 431, p. 147-151.

Ayliffe, L.K., Turner, G., and Burnard, P.G., (1993). Noble gas contents of speleothem inclusion fluids: potential as indicators of precipitation temperature. *Terra Nova*, v. 5, p. 646.

Badertscher, S., (2007). Charakterisierung von Einschlüssen in Stalagmiten zur Bestimmung der Paläotemperatur. *Unpublished diploma thesis*, ETH Zürich.

Badertscher, S.V., Scheidegger, Y., Leuenberger, M., Fleitmann, D., Wieler, R., and Kipfer, R., (2007). Trace gas content in air inclusions in speleothems as a new paleoenvironmental archive? *European Geosciences Union Abstract*, Vienna.

Baker, A., and Genty, A., (1998). Environmental pressures on conserving cave speleothems: effects of changing surface land use and increased cave tourism. *Journal of Environmental Management*, v. 53, p. 165-175.

Baldini, J.U.L., (2010). Cave atmosphere controls on stalagmite growth rate and palaeclimate records, in Pedley, H.M., and Rogerson, M., eds., *Tufas and Speleothems, Unraveling the Microbial and Physical Controls*, Volume 336, Geological Society.

Baldini, J.U.L., McDermott, F., Hoffmann, D.L., Richards, D.A., and Clipson, N., (2008). Very high-frequency and seasonal cave atmosphere PCO2 variability: Implications for stalagmite growth and oxygen isotope-based paleoclimate records. *Earth and Planetary Science Letters*, v. 272, p. 118-129.

Ballentine, C.J., and Hall, C.M., (1999). Determining paleotemperature and other variables by using an error-weighted, nonlinear inversion of noble gas concentrations in water. *Geochimica Et Cosmochimica Acta*, v. 63, p. 2315-2336.

Banner, J.L., Guilfoyle, A., James, E.W., Stern, L.A., and Musgrove, M., (2007). Seasonal variations in modern speleothem calcite growth in Central Texas, USA. *Journal of Sedimentary Research*, v. 77, p. 615-622.

Beyerle, U., Aeschbach-Hertig, W., Imboden, D.M., Baur, H., Graf, T., and Kipfer, R., (2000). A mass spectrometric system for the analysis of noble gases and tritium from water samples. *Environmental Science & Technology*, v. 34, p. 2042-2050.

Beyerle, U., Purtschert, R., Aeschbach-Hertig, W., Imboden, D.M., Loosli, H.H., Wieler, R., and Kipfer, R., (1998). Climate and groundwater recharge during the last glaciation in an ice-covered region. *Science*, v. 282, p. 731-734.

Blyth, A.J., Asrat, A., Baker, A., Gulliver, P., Leng, M.J., and Genty, D., (2007). A new approach to detecting vegetation and land-use change using high-resolution lipid biomarker records in stalagmites. *Quaternary Research*, v. 68, p. 314-324.

Blyth, A.J., and Watson, J.S., (2009). Thermochemolysis of organic matter preserved in stalagmites: A preliminary study. *Organic Geochemistry*, v. 40, p. 1029-1031.

Blyth, A.J., Watson, J.S., Woodhead, J., and Hellstrom, J., (2010). Organic compounds preserved in a 2.9 million year old stalagmite from the Nullarbor Plain, Australia. *Chemical Geology*, v. 279, p. 101-105.

Bottinga, Y., (1969). Calculated fractionation factors for carbon and hydrogen isotope exchange in system calcite-carbon dioxide – graphite – methane – hydrogen – water. *Geochimica Et Cosmochimica Acta*, v. 33, p. 49-&.

Bowen, G.J., and Revenaugh, J., (2003). Interpolating the isotopic composition of modern meteoric precipitation. *Water Resources Research*, v. 39.

Bowen, G.J., Wassenaar, L.I., and Hobson, K.A., (2005). Global application of stable hydrogen and oxygen isotopes to wildlife forensics. *Oecologia*, v. 143, p. 337-348.

Brennwald, M.S., Kipfer, R., and Imboden, D.M., (2005). Release of gas bubbles from lake sediment traced by noble gas isotopes in the sediment pore water. *Earth and Planetary Science Letters*, v. 235, p. 31-44.

Brennwald, M.S., Peeters, F., Imboden, D.M., Giralt, S., Hofer, M., Livingstone, D.M., Klump, S., Strassmann, K., and Kipfer, R., (2004). Atmospheric noble gases in lake sediment pore water as proxies for environmental change. *Geophys. Res. Lett.*, v. 31.

Buhl, D., Immenhauser, A., Smeulders, G., Kabiri, L., and Richter, D.K., (2007). Time series $\delta 26Mg$ analysis in speleothem calcite: Kinetic versus equilibrium fractionation, comparison with other proxies and implications for palaeoclimate research. *Chemical Geology*, v. 244, p. 715-729.

Burns, S.J., Fleitmann, D., Matter, A., Kramers, J., and Al-Subbary, A.A., (2003). Indian Ocean climate and an absolute chronology over Dansgaard/Oeschger events 9 to 13. *Science*, v. 301, p. 1365-1367.

Burns, S.J., Fleitmann, D., Matter, A., Neff, U., and Mangini, A., (2001). Speleothem evidence from Oman for continental pluvial events during interglacial periods. *Geology*, v. 29, p. 623-626.

Cheng, H., Edwards, R.L., Broecker, W.S., Denton, G.H., Kong, X.G., Wang, Y.J., Zhang, R., and Wang, X.F., (2009). Ice Age Terminations. *Science*, v. 326, p. 248-252.

Clever, H., L., (1979). Krypton, xenon and radon - gas solubilities, *Solubility data series*, Volume 2, Pergamon Press, Oxford.

Copeland, P., Watson, E.B., Urizar, S.C., Patterson, D., and Lapen, T.J., (2007). Alpha thermochronology of carbonates. *Geochimica Et Cosmochimica Acta*, v. 71, p. 4488-4511.

Cosford, J., Qing, H.R., Mattey, D., Eglington, B., and Zhang, M.L., (2009). Climatic and local effects on stalagmite delta C-13 values at Lianhua Cave, China. *Palaeogeography Palaeoclimatology Palaeoecology*, v. 280, p. 235-244.

Craig, H., (1965). Measurement of oxygen isotope paleotemperatures, in Tongiorgi, E., ed., *Stable Isotopes in Oceanographic Studies and Paleotemperatures*.

Dansgaard, W., (1964). Stable isotope in precipitation. *Tellus*, v. 16, p. 436-468.

Darling, W., Bath, A.H., Gibson, J.J., and Rozanski, K., (2005). Isotopes in water, in Leng, M.J., ed., *Isotopes in Paleoenvironmental Research*, Volume 10, Springer.

Dennis, P.F., Rowe, P.J., and Atkinson, T.C., (2001). The recovery and isotopic measurement of water from fluid inclusions in speleothems. *Geochimica Et Cosmochimica Acta*, v. 65, p. 871-884.

Dorale, J.A., Edwards, R.L., Ito, E., and Gonzalez, L.A., (1998). Climate and vegetation history of the midcontinent from 75 to 25 ka: A speleothem record from Crevice Cave, Missouri, USA. *Science*, v. 282, p. 1871-1874.

Dreybrodt, W., (1980). Deposition of Calcite from Thin-Films of Natural Calcareous Solutions and the Growth of Speleothems. *Chemical Geology*, v. 29, p. 89-105.

Du, Z., Allan, N.L., Blundy, J.D., Purton, J.A., and Brooker, R.A., (2008). Atomistic simulation of the mechanisms of noble gas incorporation in minerals. *Geochimica Et Cosmochimica Acta*, v. 72, p. 554-573.

Fairchild, I.J., Loader, N.J., Wynn, P.M., Frisia, S., Thomas, P.A., Lageard, J.G.A., De Momi, A., Hartland, A., Borsato, A., La Porta, N., and Susini, J., (2009). Sulfur Fixation in Wood Mapped by Synchrotron X-ray Studies: Implications for Environmental Archives. *Environmental Science & Technology*, v. 43, p. 1310-1315.

Fairchild, I.J., Smith, C.L., Baker, A., Fuller, L., Spötl, C., Mattey, D., McDermott, F., (2006). Modification and preservation of environmental signals in speleothems. *Earth-Science Reviews*, 75, 105-153.

Fairchild, I.J., and Treble, P.C., (2009). Trace elements in speleothems as recorders of environmental change. *Quaternary Science Reviews*, v. 28, p. 449-468.

Faure, G., and Mensing, T.M., (2005). *Isotopes, Principles and Applications*, Wiley.

Faust, G.T., (1950). Thermal analysis studies on carbonates, aragonite and calcite. *American Mineralogist*, v. 35, p. 207-224.

Fleitmann, D., Burns, S.J., Mangini, A., Mudelsee, M., Kramers, J., Villa, I., Neff, U., Al-Subbary, A.A., Buettner, A., Hippler, D., and Matter, A., (2007). Holocene ITCZ and Indian monsoon dynamics recorded in stalagmites from Oman and Yemen (Socotra). *Quaternary Science Reviews*, v. 26, p. 170-188.

Fleitmann, D., Burns, S.J., Mudelsee, M., Neff, U., Kramers, J., Mangini, A., and Matter, A., (2003a), Holocene forcing of the Indian monsoon recorded in a stalagmite from Southern Oman. *Science*, v. 300, p. 1737-1739.

Fleitmann, D., Burns, S.J., Neff, U., Mangini, A., and Matter, A., (2003b), Changing moisture sources over the last 330,000 years in Northern Oman from fluid-inclusion evidence in speleothems. *Quaternary Research*, v. 60, p. 223-232.

Fleitmann, D., Burns, S.J., Neff, U., Mudelsee, M., Mangini, A., and Matter, A., (2004). Palaeoclimatic interpretation of high-resolution oxygen isotope profiles derived from annually laminated speleothems from Southern Oman. *Quaternary Science Reviews*, v. 23, p. 935-945.

Fleitmann, D., Cheng, H., Badertscher, S., Edwards, R.L., Mudelsee, M., Gokturk, O.M., Fankhauser, A., Pickering, R., Raible, C.C., Matter, A., Kramers, J., and Tuysuz, O., (2009). Timing and climatic impact of Greenland interstadials recorded in stalagmites from northern Turkey. *Geophysical Research Letters*, v. 36.

Genty, D., Blamart, D., Ouahdi, R., Gilmour, M., Baker, A., Jouzel, J., and Van-Exter, S., (2003). Precise dating of Dansgaard-Oeschger climate oscillations in western Europe from stalagmite data. *Nature*, v. 421, p. 833-837.

Giannesini, S., Prinzhofer, A., Moreira, M., and Magnier, C., (2008). Influence of the bound water on molecular migration of CO_2 and noble gases in clay media. *Geochimica Et Cosmochimica Act*a, supplement, v. 72, p. A308.

Goldstein, R.H., and Reynolds, T., J., (1994) Systematics of Fluid Inclusions in Diagenetic Minerals, *Society for Sedimentary Geology short course*, 31.

Griffiths, M.L., Drysdale, R.N., Gagan, M.K., Frisia, S., Zhao, J.X., Ayliffe, L.K., Hantoro, W.S., Hellstrom, J.C., Fischer, M.J., Feng, Y.X., and Suwargadi, B.W., (2010). Evidence for Holocene changes in Australian-Indonesian monsoon rainfall from stalagmite trace element and stable isotope ratios. *Earth and Planetary Science Letters*, v. 292, p. 27-38.

Griffiths, M.L., Drysdale, R.N., Vonhof, H.B., Gagan, M.K., Zhao, J.-x., Ayliffe, L.K., Hantoro, W.S., Hellstrom, J.C., Cartwright, I., Frisia, S., and Suwargadi, B.W., (2010). Younger Dryas-Holocene temperature and rainfall history of southern Indonesia from $\delta^{18}O$ in speleothem calcite and fluid inclusions. *Earth and Planetary Science Letters*, v. 295, p. 30-36.

Hall, C.M., and Ballentine, C.J., (1996). A rigorous mathematical method for calculating paleotemperature, excess air, and paleo-salinity from noble gas concentrations in groundwater. *American Geophysical Union Abstract*, Fall Meeting, p. F178.

Harmon R.S., S., H.P., O'Neil, J.R., (1979). D/H ratios in speleothem fluid inclusions: a guide to variations in the isotopic composition of meteoric precipitation. *Earth and Planetary Science Letters*, v. 42, p. 254-266.

Heaton, T.H.E., and Vogel, J.C., (1981). Excess air in groundwater. *Journal of Hydrology*, v. 50, p. 201-216.

Henderson, G.M., (2006). Climate - Caving in to new chronologies. *Science*, v. 313, p. 620-622.

Hendy, C.H., (1971). The isotopic geochemistry of speleothems. The calculation of the effects of different modes of formation on the isotopic composition of speleothems and their applicability as paleoclimatic indicators. *Geochimica et Cosmochimica Acta*, v. 35, p. 801-824.

Hofer, M., Peeters, F., Aeschbach-Hertig, W., Brennwald, M., Holocher, J., Livingstone, D.M., Romanovski, V., and Kipfer, R., (2002). Rapid deep-water renewal in Lake Issyk-Kul (Kyrgyzstan) indicated by transient tracers. *Limnology and Oceanography*, v. 47, p. 1210-1216.

Holocher, J., Peeters, F., Aeschbach-Hertig, W., Kinzelbach, W., and Kipfer, R., (2003). Kinetic model of gas bubble dissolution in groundwater and its implications for the dissolved gas composition. *Environmental Science & Technology*, v. 37, p. 1337-1343.

Holzner, C.P., McGinnis, D.F., Schubert, C.J., Kipfer, R., and Imboden, D.M., (2008). Noble gas anomalies related to high-intensity methane gas seeps in the Black Sea. *Earth and Planetary Science Letters*, v. 265, p. 396-409.

Huber, C., Beyerle, U., Leuenberger, M., Schwander, J., Kipfer, R., Spahni, R., Severinghaus, J.P., and Weiler, K., (2006). Evidence for molecular size dependent gas fractionation in firn air derived from noble gases, oxygen, and nitrogen measurements. *Earth and Planetary Science Letters*, v. 243, p. 61-73.

Ivanovich, M., and Harmon, R.S., (1993). Uranium Series Disequilibrium: Applications to Environmental Problems.

Jouzel, J., Masson-Delmotte, V., Cattani, O., Dreyfus, G., Falourd, S., Hoffmann, G., Minster, B., Nouet, J., Barnola, J.M., Chappellaz, J., Fischer, H., Gallet, J.C., Johnsen, S., Leuenberger, M., Loulergue, L., Luethi, D., Oerter, H., Parrenin, F., Raisbeck, G., Raynaud, D., Schilt, A., Schwander, J., Selmo, E., Souchez, R., Spahni, R., Stauffer, B., Steffensen, J.P., Stenni, B., Stocker, T.F., Tison, J.L., Werner, M., and Wolff, E.W., (2007). Orbital and millennial Antarctic climate variability over the past 800,000 years. *Science*, v. 317, p. 793-796.

Jung, S.J.A., Davies, G.R., Ganssen, G.M., and Kroon, D., (2004). Synchronous Holocene sea surface temperature and rainfall variations in the Asian monsoon system. *Quaternary Science Reviews*, v. 23, p. 2207-2218.

Kendall, A.C., and Broughton, P.L., (1978). Origin of Fabrics in Speleothems Composed of Columnar Calcite Crystals. *Journal of Sedimentary Petrology*, v. 48, p. 519-538.

Kim, S.T., and Oneil, J.R., (1997). Equilibrium and nonequilibrium oxygen isotope effects in synthetic carbonates. *Geochimica Et Cosmochimica Acta*, v. 61, p. 3461-3475.

Kipfer, R., Aeschbach-Hertig, W., Peeters, F., and Stute, M., (2002). Noble gases in Lakes and Groundwaters, in Porcelli, D., Ballentine, C.J., and Wieler, R., eds., *Noble gases in Geochemistry and Cosmochemistry*, Volume 47, Mineralogical Society of America.

Kluge, T., (2008). Fluid inclusions in speleothems as a new archive for the noble gas palaeothermometer, *phD Thesis*, Heidelberg.

Kluge, T., Marx, T., Scholz, D., Niggemann, S., Mangini, A., and Aeschbach-Hertig, W., (2008). A new tool for palaeoclimate reconstruction: Noble gas temperatures from fluid inclusions in speleothems. *Earth and Planetary Science Letters*, v. 269, p. 407-414.

Klump, S., Grundl, T., Purtschert, R., and Kipfer, R., (2008). Groundwater and climate dynamics derived from noble gas, C-14, and stable isotope data. *Geology*, v. 36, p. 395-398.

Klump, S., Tomonaga, Y., Kienzler, P., Kinzelbach, W., Baumann, T., Imboden, D.M., and Kipfer, R., (2007). Field experiments yield new insights into gas exchange and excess air formation in natural porous media. *Geochimica Et Cosmochimica Acta*, v. 71, p. 1385-1397.

Kowalczk, A.J., and Froelich, P.N., (2010). Cave air ventilation and CO2 outgassing by radon-222 modeling: How fast do caves breathe? *Earth and Planetary Science Letters*, v. 289, p. 209-219.

Kreuzer, A.M., von Rohden, C., Friedrich, R., Chen, Z.Y., Shi, J.S., Hajdas, I., Kipfer, R., and Aeschbach-Hertig, W., (2009). A record of temperature and monsoon intensity over the past 40 kyr from groundwater in the North China Plain. *Chemical Geology*, v. 259, p. 168-180.

Kruger, Y., Fleitmann, D., and Frenz, M., (2008). Paleotemperatures from fluid inclusion liquid-vapor homogenization in speleothems. *Pages News*, v. 16, p. 13-14.

Kruger, Y., Stoller, P., Ricka, J., and Frenz, M., (2007). Femtosecond lasers in fluid-inclusion analysis: overcoming metastable phase states. *European Journal of Mineralogy*, v. 19, p. 693-706.

Lachniet, M.S., (2009). Climatic and environmental controls on speleothem oxygen-isotope values. *Quaternary Science Reviews*, v. 28, p. 412-432.

Lecuyer, C., and Oneil, J.R., (1994). Stable-isotope compositions of fluid inclusions in biogenic carbonates. *Bulletin De La Societe Geologique De France*, v. 165, p. 573-581.

Leuenberger, M., Nyfeler, P., Moret, H.P., Sturm, P., and Huber, C., (2000a). A new gas inlet system for an isotope ratio mass spectrometer improves reproducibility. *Rapid Communications in Mass Spectrometry*, v. 14, p. 1543-1551.

Leuenberger, M., Nyfeler, P., Moret, H.P., Sturm, P., Indermuhle, A., and Schwander, J., (2000b). CO_2 concentration measurements on air samples by mass spectrometry. *Rapid Communications in Mass Spectrometry*, v. 14, p. 1552-1557.

Leuenberger, M.C., Eyer, M., Nyfeler, P., Stauffer, B., and Stocker, T.F., (2003). High-resolution delta C-13 measurements on ancient air extracted from less than 10cm^3 of ice. *Tellus Series B-Chemical and Physical Meteorology*, v. 55, p. 138-144.

Liu, D.B., Wang, Y.J., Cheng, H., Edwards, R.L., Kong, X.G., Wang, X.F., Hardt, B., Wu, J.Y., Chen, S.T., Jiang, X.Y., He, Y.Q., Dong, J.G., and Zhao, K., (2010). Sub-millennial variability of Asian monsoon intensity during the early MIS 3 and its analogue to the ice age terminations. *Quaternary Science Reviews*, v. 29, p. 1107-1115.

Ma, L., Castro, M.C., and Hall, C.M., (2004). A late Pleistocene-Holocene noble gas paleotemperature record in southern Michigan. *Geophysical Research Letters*, v. 31.

Mann, S., (2001). Biomineralization : principles and concepts in bioinorganic materials chemistry, Oxford University Press.

Mayewski, P.A., Rohling, E.E., Curt Stager, J., KarlÈn, W., Maasch, K.A., David Meeker, L., Meyerson, E.A., Gasse, F., van Kreveld, S., Holmgren, K., Lee-Thorp, J., Rosqvist, G., Rack, F., Staubwasser, M., Schneider, R.R., and Steig, E.J., (2004). Holocene climate variability. *Quaternary Research*, v. 62, p. 243-255.

McDermott, F., (2004). Palaeo-climate reconstruction from stable isotope variations in speleothems: a review. *Quaternary Science Reviews*, v. 23, p. 901-918.

McDermott, F., Schwarcz, H., Rowe, P.J., (2005). Isotopes in Speleothems, in Leng, M.J., ed., *Isotopes in Paleoenvironmental Research*, Volume 10, Springer.

McGarry, S., Bar-Matthews, M., Matthews, A., Vaks, A., Schilman, B., and Ayalon, A., (2004). Constraints on hydrological and paleotemperature variations in the Eastern Mediterranean region in the last 140 ka given by the delta D values of speleothem fluid inclusions. *Quaternary Science Reviews*, v. 23, p. 919-934.

Mercury, L., Pinti, D.L., and Zeyen, H., (2004). The effect of the negative pressure of capillary water on atmospheric noble gas solubility in ground water and palaeotemperature reconstruction. *Earth and Planetary Science Letters*, v. 223, p. 147-161.

Mohapatra, R.K., Schwenzer, S.P., Herrmann, S., Murty, S.V.S., Ott, U., and Gilmour, J.D., (2009). Noble gases and nitrogen in Martian meteorites Dar al Gani 476, Sayh al Uhaymir 005 and Lewis Cliff 88516: EFA and extra neon. *Geochimica Et Cosmochimica Acta*, v. 73, p. 1505-1522.

Momma, K., and Izumi, F., (2008). VESTA: a three-dimensional visualization system for electronic and structural analysis. *Journal of Aplied Crystallography*, v. 41, p. 653-658.

Oneil, J.R., and Adami, L.H., (1969). Oxygen isotope partition function ratio of water and structure of liquid water. *Journal of Physical Chemistry*, v. 73, p. 1553-&.

Ozima, M., and Podosek, F.A., (2002). Noble gas geochemistry, Cambridge University Press.

Pearson, F.J., Arcos, D., Bath, A.H., Boisson, Y.J., Fernandez, H.E., Gaucher, E., Gautschi, A., Griffault, L., Hernan, P., and Waber, H.N., (2003). Mont Terri Project - Geochemistry of water in the opalinus clay formation at the Mont Terri rock laboratory, *Federal office for water and geology*, Switzerland.

Perrette, Y., Delannoy, J.J., Desmet, M., Lignier, V., and Destombes, J.L., (2005). Speleothem organic matter content imaging. The use of a Fluorescence Index to characterise the maximum emission wavelength. *Chemical Geology*, v. 214, p. 193-208.

Petit, J.R., Jouzel, J., Raynaud, D., Barkov, N.I., Barnola, J.M., Basile, I., Bender, M., Chappellaz, J., Davis, M., Delaygue, G., Delmotte, M., Kotlyakov, V.M., Legrand, M., Lipenkov, V.Y., Lorius, C., Pepin, L., Ritz, C., Saltzman, E., and Stievenard, M., (1999). Climate and atmospheric history of the past 420,000 years from the Vostok ice core, Antarctica. *Nature*, v. 399, p. 429-436.

Pettitt, A.N., (1979). A non-parametric approach to the change-point problem. *Applied Statistics*, v. 28, p. 126-135.

Pons-Branchu, E., Hamelin, B., Losson, B., Jaillet, S., and Brulhet, J., (2010). Speleothem evidence of warm episodes in northeast France during Marine Oxygen Isotope Stage 3 and implications for permafrost distribution in northern Europe. *Quaternary Research*, v. 74, p. 246-251.

Porcelli, D., Ballentine, C.J., and Wieler, R., (2002). An overview of noble gas geochemistry and cosmochemistry, in Porcelli, D., Ballentine, C.J., and Wieler, R., eds., *Noble gases in Geochemistry and Cosmochemistry*, Volume 47, Mineralogical Society of America.

Poulson, T.L., and White, W.B., (1969). Cave environment. *Science*, v. 165, p. 971-981-

Rodionov, S.N., (2004). A sequential algorithm for testing climate regime shifts. *Geophysical Research Letters*, v. 31.

Roedder, E., (1984). Fluid Inclusions. *Reviews in Mineralogy*, v. 12.

Romanek, C.S., Grossman, E.L., and Morse, J.W., (1992). Carbon isotopic fractionation in synthetic aragonite and calcite – effects of temperature and precipitation rate. *Geochimica Et Cosmochimica Acta*, v. 56, p. 419-430.

Rozanski, K., Aragues-Araguas, L., and Gonfiantini, R., (1993). Isotopic patterns in modern global precipitation, in Swart, P.K., Lohmann, K.C., McKenzie, J., and Savin, S., eds., *Climate change in continental isotopic records*, Washington DC, American Geophysical Union, p. 1-36.

Scheidegger, Y., Badertscher, S., Wieler, R., Heber, V., and Kipfer, R., (2007a). Microscopical speleothem calcite investigations proofing the existence of two different types of fluid inclusions, *European Geosciences Union abstract*: Vienna.

Scheidegger, Y., Baur, H., Brennwald, M.S., Fleitmann, D., Wieler, R., and Kipfer, R., (2010). Accurate analysis of noble gas concentrations in small water samples and its application to fluid inclusions in stalagmites. *Chemical Geology*, v. 272, p. 31-39.

Scheidegger, Y., Wieler, R., Brennwald, M.S., Fleitmann, D., Jeannin, P.Y. and Kipfer, R., (2010). Determination of Holocene cave temperatures from Kr and Xe concentrations in stalagmite fluid inclusions. *Submitted for publication in Chemical Geology, currently in review*.

Scheidegger, Y., Kipfer, R., Wieler, R., Badertscher, S., and Leuenberger, M., (2007b). Noble gases in fluid inclusions in speleothems. *Geochimica Et Cosmochimica Acta*, v. 71, p. A886-A886.

Scholte, P., and De Geest, P., (2010). The climate of Socotra Island (Yemen): A first-time assessment of the timing of the monsoon wind reversal and its influence on precipitation and vegetation patterns. *Journal of Arid Environments*, v. 74, p. 1507-1515.

Schwarcz, H.P., Harmon, R.S., (1976). Stable isotope studies of fluid inclusions in speleothems and their paleoclimatic significance. *Geochimica et Cosmochimica Acta*, v. 40, p. 657-665.

Shakun, J.D., Burns, S.J., Fleitmann, D., Kramers, J., Matter, A., and Al-Subary, A., (2007). A high-resolution, absolute-dated deglacial speleothem record of Indian Ocean climate from Socotra Island, Yemen. *Earth and Planetary Science Letters*, v. 259, p. 442-456.

Smithson, P.A., (1991). Interrelationships between cave and outside air temperatures. *Theoretical and Applied Climatology*, v. 44, p. 65-73.

Spotl, C., Fairchild, I.J., and Tooth, A.F., (2005). Cave air control on dripwater geochemistry, Obir Caves (Austria): Implications for speleothem deposition in dynamically ventilated caves. *Geochimica Et Cosmochimica Acta*, v. 69, p. 2451-2468.

Spotl, C., and Mangini, A., (2002). Stalagmite from the Austrian Alps reveals Dansgaard-Oeschger events during isotope stage 3: Implications for the absolute chronology of Greenland ice cores. *Earth and Planetary Science Letters*, v. 203, p. 507-518.

Strassmann, K.M., Brennwald, M.S., Peeters, F., and Kipfer, R., (2005). Dissolved noble gases in the porewater of lacustrine sediments as palaeolimnological proxies. *Geochimica Et Cosmochimica Acta*, v. 69, p. 1665-1674.

Stute, M., Clark, J.F., Schlosser, P., Broecker, W.S., and Bonani, G., (1995a). A 30,000-yr contitental paleotemperature record derived from noble gases dissolved in groundwater from the san juan basin, new-mexico. *Quaternary Research*, v. 43, p. 209-220.

Stute, M., Forster, M., Frischkorn, H., Serejo, A., Clark, J.F., Schlosser, P., Broecker, W.S., and Bonani, G., (1995b), Cooling of tropical Brazil (5°C) duing the last glacial maximum. *Science*, v. 269, p. 379-383.

Stute, M., and Schlosser, P., (2000). Environmental tracers in subsurface hydrology, in Herczeg, A.L., ed., *Atmospheric Noble gases*: Boston, Kluwer Academic Publishers, p. 349–377.

Stute, M., Schlosser, P., Clark, J.F., and Broecker, W.S., (1992). Paleotemperatures in the southwestern united states derived from noble gases in groundwater. *Science*, v. 256, p. 1000-1003.

Thomas, J.B., Cherniak, D.J., and Watson, E.B., (2008). Lattice diffusion and solubility of argon in forsterite, enstatite, quartz and corundum. *Chemical Geology*, v. 253, p. 1-22.

Treble, P.C., Chappell, J., and Shelley, J.M.G., (2005). Complex speleothem growth processes revealed by trace element mapping and scanning electron microscopy of annual layers. *Geochimica Et Cosmochimica Acta*, v. 69, p. 4855-4863.

Vaks, A., Bar-Matthews, M., Matthews, A., Ayalon, A., and Frumkin, A., (2010). Middle-Late Quaternary paleoclimate of northern margins of the Saharan-Arabian Desert: reconstruction from speleothems of Negev Desert, Israel. *Quaternary Science Reviews*, v. 29, p. 2647-2662.

van Breukelen, M.R., Vonhof, H.B., Hellstrom, J.C., Wester, W.C.G., and Kroon, D., (2008). Fossil dripwater in stalagmites reveals Holocene temperature and rainfall variation in Amazonia. *Earth and Planetary Science Letters*, v. 275, p. 54-60.

Wainer, K., Genty, D., Blamart, D., Daîron, M., Bar-Matthews, M., Vonhof, H., Dublyansky, Y., Pons-Branchu, E., Thomas, L., van Calsteren, P., Quinif, Y., and Caillon, N., Speleothem record of the last 180ka in Villars cave (SW France): Investigation of a large $\delta^{18}O$ shift between MIS6 and MIS5. *Quaternary Science Reviews*, v. 30, p. 130-146.

Wang, Y.J., Cheng, H., Edwards, R.L., Kong, X.G., Shao, X.H., Chen, S.T., Wu, J.Y., Jiang, X.Y., Wang, X.F., and An, Z.S., (2008). Millennial- and orbital-scale changes in the East Asian monsoon over the past 224,000 years. *Nature*, v. 451, p. 1090-1093.

Wanner, H., Beer, J.r., B̧tikofer, J., Crowley, T.J., Cubasch, U., Fļckiger, J., Goosse, H., Grosjean, M., Joos, F., Kaplan, J.O., Ķttel, M., M̧ller, S.A., Prentice, I.C., Solomina, O., Stocker, T.F., Tarasov, P., Wagner, M., and Widmann, M., (2008). Mid- to Late Holocene climate change: an overview. *Quaternary Science Reviews*, v. 27, p. 1791-1828.

Weiss, R.F., (1970). Solubility of nitrogen, oxygen and argon in water and seawater. *Deep-Sea Research*, v. 17.

Weiss, R.F., (1971). Solubility of helium and nein in water and seawater. *Journal of Chemical and Engineering Data*, v. 16, p. 235-241, p. 721-735.

Weiss, R.F., and Kyser, T.K., (1978). Solubility of krypton in water and seawater. *Journal of Chemical and Engineering Data*, v. 23, p. 69-72.

Weyhenmeyer, C.E., Burns, S.J., Waber, H.N., Aeschbach-Hertig, W., Kipfer, R., Loosli, H.H., and Matter, A., (2000). Cool glacial temperatures and changes in moisture source recorded in Oman groundwaters. *Science*, v. 287, p. 842-845.

Wynn, P.M., Fairchild, I.J., Frisia, S., Spotl, C., Baker, A., and Borsato, A., (2010). High-resolution sulphur isotope analysis of speleothem carbonate by secondary ionisation mass spectrometry. *Chemical Geology*, v. 271, p. 101-107.

Zhang, R., Schwarcz, H.P., Ford, D.C., Schroeder, F.S., and Beddows, P.A., (2008). An absolute paleotemperature record from 10 to 6 Ka inferred from fluid inclusion D/H ratios of a stalagmite from Vancouver Island, British Columbia, Canada. *Geochimica Et Cosmochimica Acta*, v. 72, p. 1014-1026.

Zhao, K., Wang, Y., Edwards, R.L., Cheng, H., and Liu, D., (2010). High-resolution stalagmite $\delta^{18}O$ records of Asian monsoon changes in central and southern China spanning the MIS 3/2 transition. *Earth and Planetary Science Letters*, v. 298, p. 191-198.

i want morebooks!

Buy your books fast and straightforward online - at one of world's fastest growing online book stores! Environmentally sound due to Print-on-Demand technologies.

Buy your books online at
www.get-morebooks.com

Kaufen Sie Ihre Bücher schnell und unkompliziert online – auf einer der am schnellsten wachsenden Buchhandelsplattformen weltweit! Dank Print-On-Demand umwelt- und ressourcenschonend produziert.

Bücher schneller online kaufen
www.morebooks.de

VDM Verlagsservicegesellschaft mbH
Heinrich-Böcking-Str. 6-8　　Telefon: +49 681 3720 174　　info@vdm-vsg.de
D - 66121 Saarbrücken　　　Telefax: +49 681 3720 1749　　www.vdm-vsg.de

Printed by Books on Demand GmbH, Norderstedt / Germany